Lyman®

Metallic Reloading Basics

C. Kenneth Ramage, editor

Metallic Reloading Basics, First Edition

Printed in the U.S.A.

ISBN: 0-912412-11-9
Library of Congress: 76-209-31

Published by Lyman Publications,
Route 147
Middlefield, Ct. 06455
J. Mace Thompson, president

INTRODUCTION

Reloading is a natural step for any shooter. Actually, reloading may be seen as a hobby separate from, but closely related to, shooting. For any person with some scientific bent and an inquisitive nature, reloading offers an opportunity to get more performance from his guns while enjoying himself—and saving a considerable chunk of money, to boot.

This first edition of Metallic Reloading Basics is intended to guide the beginner while providing some data on popular bullet weights within the most popular calibers. For advanced information and a much wider array of reloading data, a copy of the Lyman Reloading Handbook, 45th Edition, is invaluable.

Welcome to reloading; we hope you find this publication useful. If you have questions or problems, our technical staff will assist—just drop them a note.

Table of Contents

1. Getting Started 5
2. Cartridge Cases 6
3. Powder 9
4. Primers 11
5. Projectiles 13
6. Reloading Operations 14
7. Reloading Data 23
8. Lyman Reloading Equipment . . 54

GETTING STARTED

In view of your present interest in the reloading subject, it is safe to assume that you have already had considerable shooting experience. Certainly, you have purchased, and fired, a good many rounds of factory loaded ammunition. Hence you are equipped with a working knowledge of the various caliber designations and cartridge shapes, and you have learned to recognize the correct cartridge for your particular rifle. Probably many readers will have an overall understanding of shooting that extends beyond these simple basics. However, for the benefit of the uninitiated, we'll begin at this point.

Each factory loaded cartridge we purchase and fire represents extensive thought and care on the part of its manufacturer. If we are to reload a fired cartridge and duplicate, if possible, the original factory loading, we must first learn to appreciate some of the intricacies of ammunition design. For instance, the fired cartridge case is the most important reloading component. The function of the case, how it is constructed, and its **condition after firing** are important considerations to reloaders.

Equipment Selection

You'll need to equip a reloading bench. While there are much less sophisticated reloading systems on the market, starting with the best gear you can afford quickly pays for itself by turning out excellent reloads on a volume basis.

We've listed our thoughts on basic equipment below and, for reloaders needing complete sets of matched equipment, Lyman now provides the two reloading kits illustrated which contain all needed items—plus an accessory collection for the reloader already owning a press.

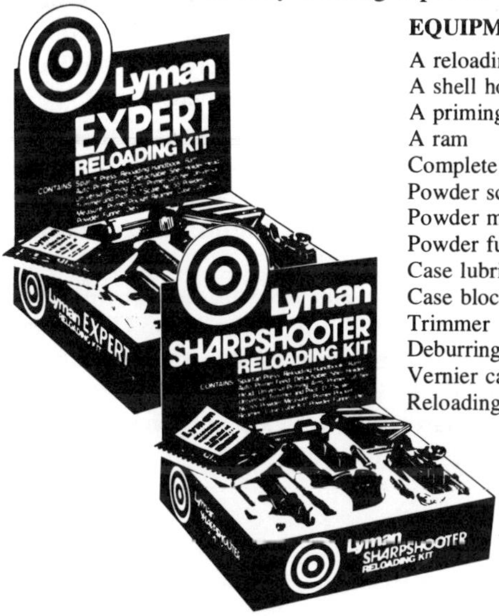

EQUIPMENT LIST:

A reloading press
A shell holder
A priming arm
A ram
Complete set of dies
Powder scale
Powder measure
Powder funnel
Case lubricant with pad
Case blocks
Trimmer
Deburring tool
Vernier caliper or case length gauge
Reloading manual

Optional Equipment:

Automatic primer feed
Case neck brush
Inside neck reamers
Outside neck turners
Drop tube funnel
Primer pocket cleaner
Primer pocket reamer
Powder Trickler
Bullet puller

CARTRIDGE CASES

The chief function of a cartridge case is to seal off the breech at the time of firing. To accomplish this, the case walls must expand freely so that they are tight against the sides of the chamber. This sealing action prevents the hot powder gases from leaking back around the cartridge and out through the action. Along with this, the cartridge case must withstand the chamber pressure that is built up during firing. To achieve this, the case requires a structural strength of its own—plus the additional supporting strength supplied by the bolt face and chamber walls. In essence, the case functions as an intrinsic part of the rifle. **A rifle is no stronger than the case that is used in it—nor is the case stronger than the rifle.**

Cartridge brass is carefully tempered in its final manufacture. The head of the case is thick and tough which gives it the strength and rigidity necessary to resist the force of the chamber pressure. The forward section of the case (neck, shoulder and body) is considerably thinner than the head section. In manufacture, these portions are given an anneal which leaves them soft and ductile. The obvious advantage is that the case walls and neck will now expand freely to release the bullet and seal the chamber while the cartridge is fired.

Cartridge Case Nomenclature

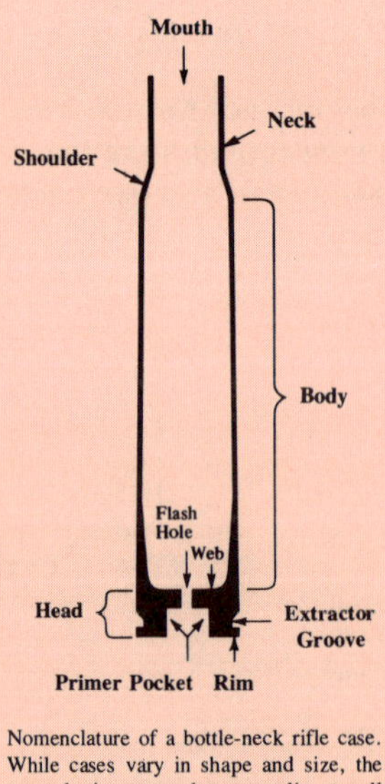

Nomenclature of a bottle-neck rifle case. While cases vary in shape and size, the same basic nomenclature applies to all center fire cartridges.

As shooters, we may have been rather casual in our regard for empty brass cases, but as reloaders we soon come to think differently. Without a quantity of strong and serviceable cases, we would not get far in reloading ammunition. The most usual way for a reloader to obtain serviceable cases is to purchase factory loaded ammunition. After this ''store bought'' ammo has been fired, the empty cases are retained for future reloading. The reloader may also purchase new cases from his component dealer.

To make sure your cases are in prime condition, start with either new or once-fired cases. Never use brass of unknown origin such as that found on a shooting range.

Each firing and resizing has an influence on the serviceability of the case. The battering of chamber pressure and the forces applied by the resizing die eventually work-harden the forward portion of the case and destroy its usefulness.

Cartridge Cases, Continued

CASE STYLES

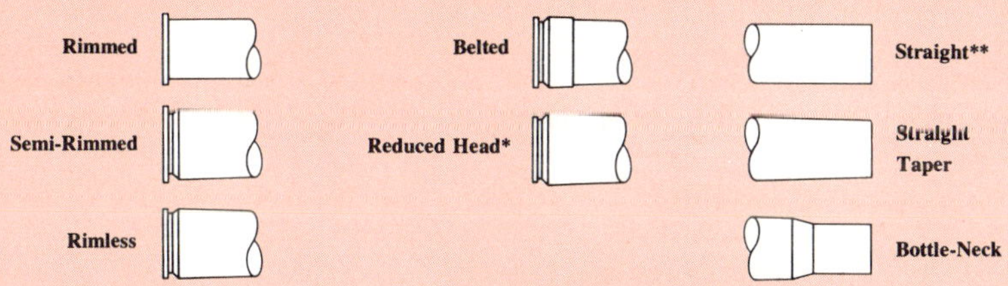

*Function is basically the same as Rimless Type.
**Straight walled cases have a slight taper which is not visibly apparent.

This illustration shows the head forms and body shapes standard to cartridge design. As you will see later, the design of the cartridge case governs the headspace dimensions. Case examples for almost any combination of the above shapes may be encountered. For example: the 30/30 uses the rimmed bottle-neck, the 30/06 uses the rimless bottle-neck, the .458 Winchester uses the belted head and the straight body.

Carefully inspect your cases before each reloading. If your cases are new, or once fired, they will not reveal fatigue at the first reloading. However, fatigue signs will show up in subsequent loading, so you must learn to look for them. **Check your cases for splits or cracks in the neck, shoulder or body.** Reject all cases that show signs of defects, but flatten them with a pair of pliers before discarding to prevent their reuse. We suggest you separate your cartridge cases into lots and keep a record of their history.

Trimming is necessary when your cases have lengthened after numerous firings. Check your cases after resizing and never allow them to exceed the maximum listed measurement. How often you will need to trim the cases depends largely on the type of case you are using and the pressure of the load. Bottle-neck cases take more abuse from pressure than straight sided cases and require trimming more often. The trimming of any case more than four times is not recommended. After this amount of trimming, it may be assumed that the case walls are now too thin, and the case should be discarded.

Case Diagnosis

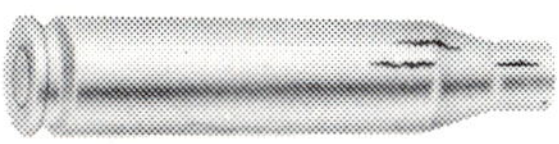

CONDITION "A"
CASE FATIGUE

The condition pictured above illustrates case fatigue. Note how the cracks run lengthwise. Sometimes only pin hole cracks are noticeable, but such cases must also be discarded. Pistol cases show similar cracks.

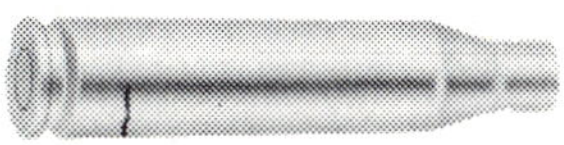

CONDITION "B"
EXCESSIVE HEAD SPACE

The condition pictured in this example is quite different from case fatigue. This crack runs around the circumference of the case and indicates an excessive headspace condition. Never use any rifle which shows signs of excessive headspace.

To properly check the length of your cases, you will require a good vernier caliper available at most hardware stores.

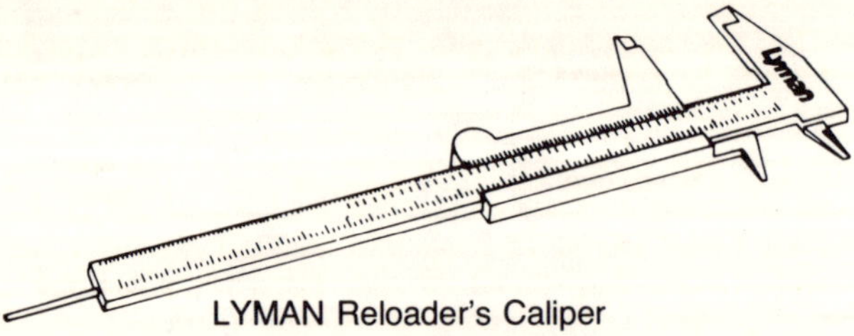

LYMAN Reloader's Caliper

A vernier caliper, 4" long or longer, is a necessary reloading tool.

In addition, the reloader needs an accurate means of case trimming. The Lyman Universal Trimmer (pictured below) is actually a miniature lathe featuring a unique chuck head which accepts all metallic cartridge cases, rifle or pistol. The tool is supplied with complete instructions and will accomplish the trimming operation quickly and easily.

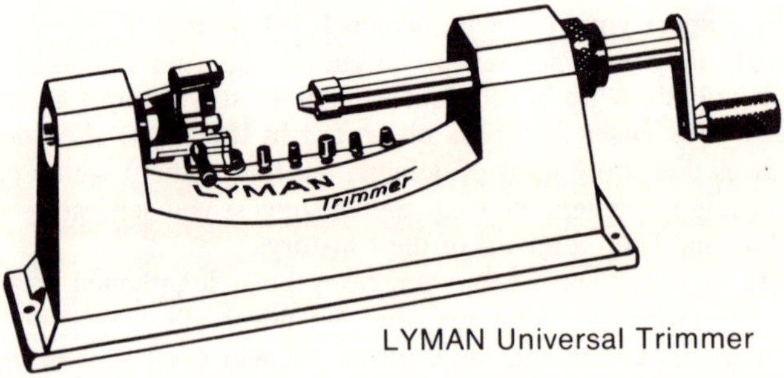

LYMAN Universal Trimmer

When loading new or once-fired cases, it is necessary to remove the sharp inside edges of the case mouth. This operation is called **chamfering** and its purpose is to ease insertion of the new bullet. Chamfering is required only for the first reloading of a new or once-fired case. The inexpensive Lyman hand reamer chamfers a case easily and with uniformity. Hold the case in one hand, while you lightly turn the reamer in the case mouth with the other hand. Remove very little material and **do not** cut a sharp knife edge on the case.

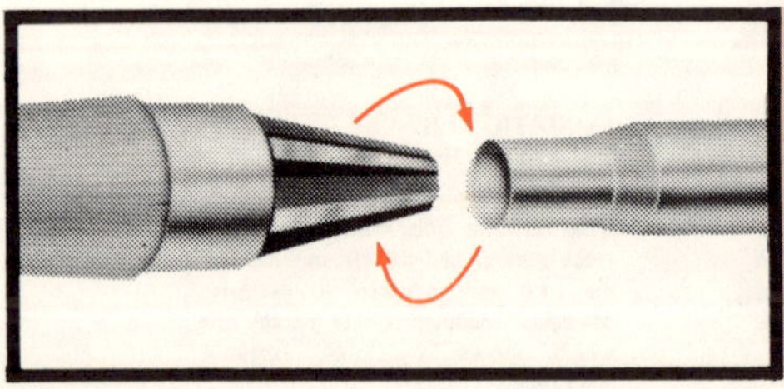

When chamfering turn the reamer lightly, removing very little case material.

POWDER

Many people are concerned with the dangers of handling and loading modern smokeless powders. To the uninitiated, gun powder is often regarded as "explosive"—something to be feared such as dynamite. This is quite a natural presumption.

The various modern powders used in reloading are classified as propellents, which means they are chemical mixtures designed to burn under controlled conditions, and to propel a shot charge or projectile. A high explosive, on the other hand, is a completely different breed of cat. These mixtures are quite dangerous because they are designed to detonate. When a substance such as dynamite or blasting gelatin, is detonated, it produces intense heat and violent shock waves. These shock waves exert tremendous pressure on anything they contact, which makes it almost impossible to vent away the effects of detonation involving any appreciable quantity of explosive material.

While not to be compared to explosives, modern reloading powders are nonetheless highly flammable. They not only burn, they burn vigorously. In case of accidental ignition, a great amount of gas at high temperature will be formed. If the powder is stored in its original factory canister, **as it should be,** this gas will create a pressure on the relatively fragile sides of the can. The pressure will split open the seams of the container, or pop off the lid. In such an event, the pressure remains at a low level **if sufficient space is provided to accommodate the escaping gas and vent it away.**

For safe reloading, large quantities of reloading powder should never be stock-piled. The few cans that you do store must be handled with the care and caution due a flammable substance. In this way you control the situation so that it cannot get out of hand. Even in the case of a total house fire arising from other sources, **smokeless powders will not produce the effect of an explosion** if the proper storage precautions have been taken.

Strong Storage Cabinet
Walls restrain the gas, allowing it to build up to a violent release.

Weak Storage Cabinet
Wall, or walls, push out to allow easy venting of gas before appreciable build up of force.

The burning characteristics of smokeless powders are complex in nature and, **depending on the application of the powder, their burning rate can change drastically.** While the chemical composition of a powder, the shape and size of its particles and the density or porosity of the powder composition, tend to control burning rate, the application of each powder must be carefully considered. Other conditions such as the degree of confinement, the heat of ignition, the temperature of the combustion chamber, the chamber pressure, and the density of loading, all affect the burning rate, yet such vital factors cannot possibly be included in the so called burning rate charts.

All reloading powders are manufactured to an exacting set of specifications. They are sold in factory sealed canisters with the name, or number, of each specific powder clearly printed on the label. When you purchase a reloading powder, you can be sure of obtaining the exact powder specified, and that the contents of the can have not been tampered with.

When weighing a powder charge, you depend on the accuracy of your powder scale and, of course, care must be taken. Reloading scales, such as the Lyman D-7, are designed specifically for the weighing of powders. These scales have a guaranteed accuracy of one-tenth of a grain, which is more than adequate for the most demanding reloading requirement.

PRIMERS

NOTE: This chapter refers only to rifle and pistol cartridge primers. For information pertaining to shotshell primers, see the Lyman Shotshell Handbook, 2nd Edition.

For a cartridge to function properly, the successful operation of each component is required. Primarily from a safety standpoint, we have stated that the cartridge case is the most important component. However, if a primer does not ignite, we will have a misfire and all of our concern for case inspection, powder selection, etc., will be for naught. Therefore, every component, including the primer, is critical to the performance of the load.

Actually, the primer is responsible for only a small percentage of the accuracy of a given load, but its burning characteristics will add or detract from the overall pressure. How much these burning characteristics affect pressure is relative. It depends primarily upon the application of a specific primer to a specific set of conditions. Generally, the primer's influence on pressure is minimal when compared to the possible effects of other conditions and components. This influence, however, is not so slight that it can be overlooked.

In testing metallic cartridges we noted pressure variations (due to primer change) of approximately 2,000 C.U.P. Further testing could indicate still greatter variations. The point is that while metallic cartridge primers do not seem to vary greatly, enough variation exists to require us to re-work a load when changing primer type or brand. The reloader should stick with the same brand and type of primer when working up and using a load. If he changes the primer brand or type, he must then go back to the "starting load" and begin again.

If you examine a "Boxer Type" primer carefully, you will note that the anvil protrudes slightly beyond the end of the primer cup. This is not an oversight in manufacture! The anvil is supposed to protrude so that it will seat solidly against the bottom of the primer pocket and firmly resist the force of the

NOMENCLATURE OF "BOXER TYPE" PRIMER

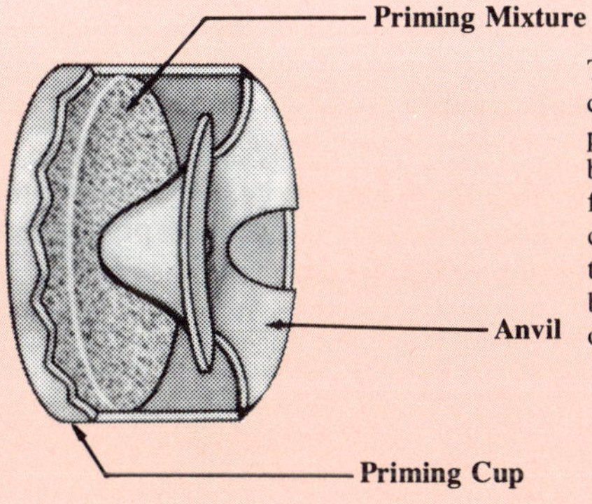

The boxer type primer is used in all American metallic cartridges. The battery cup primer used in shotshells is of the same basic design, but contains an extra cup to facilitate its use in the less rigid shotshell case. The antique Berdan type primer functions in the same manner as the Boxer type, but with this primer the anvil is actually part of the case.

firing pin blow. If the primer is seated incorrectly (leaving space between anvil and bottom of pocket) then the entire primer will move forward when struck by the firing pin. Such a condition retards the blow of the firing pin and causes inconsistent ignition.

When seating primers, make sure that the primer bottoms in the primer pocket. However, the primer should not be crushed. Normally, when a primer is seated correctly, it will be flush with the head of the case, or perhaps a few thousandths below. Under no circumstances should the primer stick out beyond the case head. Such a condition can prove dangerous particularly in auto loading firearms, for the primer is in an exposed position. If such an improperly seated primer were to receive a blow from the bolt face, it could fire before the round was safely chambered.

Boxer primers having either round or flat dome construction may be encountered and the correctly shaped priming punch is available for either contour. Use the flat punch for the flat primer and the concave punch for the round primer. Before seating a primer it is wise to examine it visually to make sure the anvil is not missing. Although this rarely occurs, the resultant misfire could be embarrassing if it should come at the wrong moment (such as when that trophy buck shows himself from behind the hemlocks).

Handling Primers

A word of caution on handling primers. Primers are detonated by percussion (a sharp blow). As packaged by the manufacturer, primers are quite safe, but they should not be tossed about or handled in a careless manner. Keep your primers in the factory container until they are to be used. Never store primers in a makeshift container. An old glass jar full of primers could become lethal if dropped or accidentally knocked onto the floor.

Crimped Primers

Some military cases are loaded in a manner which employs a slight crimp over the primer to hold it snugly in place. These cases can be decapped without difficulty, but the crimp must be removed before the case can be primed again. Various commercial tools, which include the Lyman primer pocket reamer, are available to do this job. Primer pocket reamers come in two sizes, large and small.

One last thought—never decap live primers. If for some reason live ammunition must be disassembled, pull the bullet, pour out the powder, chamber the primed case in a firearm, and snap the primer before decapping.

PROJECTILES

Today's metallic reloader uses, primarily, jacketed bullets. These copper-clad projectiles are far better performers, in all but a very few applications, than are bullets cast of lead alloy. Since jacketed bullets are the most popular, we'll limit our discussion to that type although the same principles also apply to cast bullets.

There are several factors to consider when selecting a bullet besides having carefully tested load data for a given projectile weight:

1. Weight—For optimum potential accuracy the bullet must fall within a range peculiar to a given caliber. Also, the velocity at which that bullet will be driven is a factor. The whole idea is to properly blend the three (weight, rate of twist and velocity) to produce the results satisfying you, the reloader. Weight is important to the target shooters for its effect upon sectional density; hunters are concerned with weight for its implications to penetration potential on game.

2. Base Design—There are two types: flat base and boat tail. Of the two, the boat tail has the most potential for accuracy in that the base can be more precisely made. Also, the boat tail will produce less aerodynamic drag than will an otherwise identical projectile with a flat base. This means the boat tail design will shoot flatter at extended range than will a flat base; all other variables being equal.

3. Point Design—There are various configurations within a given caliber ranging from aerodynamically sleek spitzer points for minimum drag to round noses particularly suited for short range hunting loads. These point shapes can also incorporate an exposed lead nose, hollow point or full coverage jacket.

One last thing to remember about jacketed bullets is that the actual jacket material and projectile bearing surface will vary between manufacturers for bullets of identical weight and similar design.

All our data lists the specific bullet for which the loads were developed.

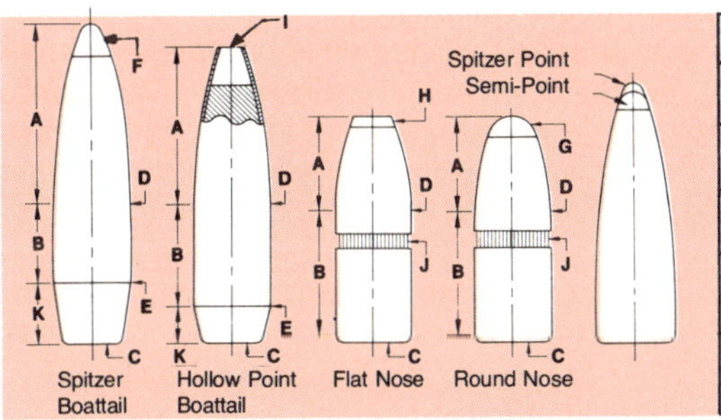

TABULATION			
A	Ogive	F	Exposed Lead (Spitzer Point)
B	Bearing Surface	G	Exposed Lead (Round Nose)
C	Base	H	Exposed Lead (Flat Nose)
D	Junction of Ogive and Bearing Surface	I	Hollow Point
		J	Cannelure
E	Junction of Boattail and Bearing Surface	K	Boattail Length

RELOADING OPERATIONS

There are only six basic mechanical operations required to reload a cartridge. Four of them are performed by the reloading dies. As you read through the text and take note of the illustrations, you will see how a set of only two reloading dies can accomplish all four of these operations.

1. **FULL LENGTH RESIZING:** When a cartridge is fired, the neck, shoulder and body of the case expand to seal the chamber releasing the bullet. Since all chambers are not identical, cases fired in one rifle may not chamber in another unless their walls are reduced to a standard diameter that is acceptable to all rifles. This operation is called resizing.
2. **DECAPPING:** This operation consists simply of removing the old primer.
3. **INSIDE NECK EXPANDING:** After a case has been resized, the inside diameter of its neck will be too small to accept the new bullet. Inside neck expanding enlarges the diameter of the neck to a size that will receive and hold the bullet securely. For pistol cases, a two-step expanding plug is used to open up the inside of the case neck. The first step on this plug is slightly smaller than bullet diameter, while the second step is a few thousandths large. The idea behind this is to allow the bullet to enter the case freely without shaving lead. The actual difference between the two steps is not visually apparent.
4. **PRIMING:** This operation consists of inserting a new primer into the primer pocket.
5. **CHARGING POWDER:** This operation consists of carefully weighing out and pouring the appropriate powder charge into the case.
6. **BULLET SEATING:** The last operation in the reloading process is the seating of a bullet into the case.

Making Cartridges

Now, if your cases have been properly inspected and you've selected a load, and purchased the necessary components (primers, powder and bullets), you are ready to begin. Your reloading press should be assembled and mounted according to the instructions supplied with the tool. Many reloading presses may be assembled to function either on the up, or on the down-stroke of the handle. The press pictured in our illustrations is operating on the down-stroke. First, lubricate your cases by wiping them with a cloth sparingly greased with Lyman Resizing Lubricant. Be frugal for too much lubricant will trap air in the die and cause "lube dents". Cases dented in this manner may be used for reloading, for the dents will be ironed out in firing. It is not considered good reloading practice, however, and care should be exercised.

Screw the Full-Length Resizing Die into the head of your press and adjust it according to the instructions supplied with the die.

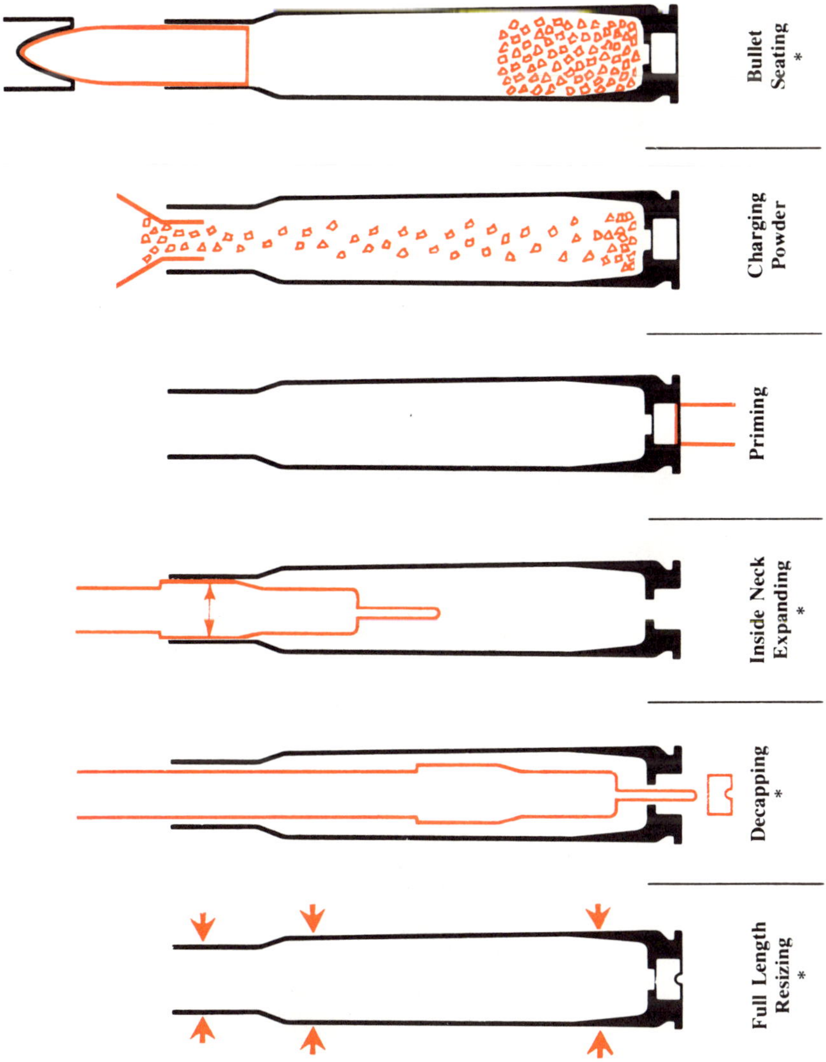

*Operations 1,2,3 and 6 are performed by reloading dies.

STEP 1

Step One *(Full-Length Resizing and Decapping)*

Slide the head of your cartridge case into the shell holder and pull the press handle down all the way. If the die is adjusted properly, the entire cartridge case will enter the die flush to the shell holder. Note in the cutaway drawing how two of the original six reloading operations (full-length resizing and decapping) are accomplished by this step.

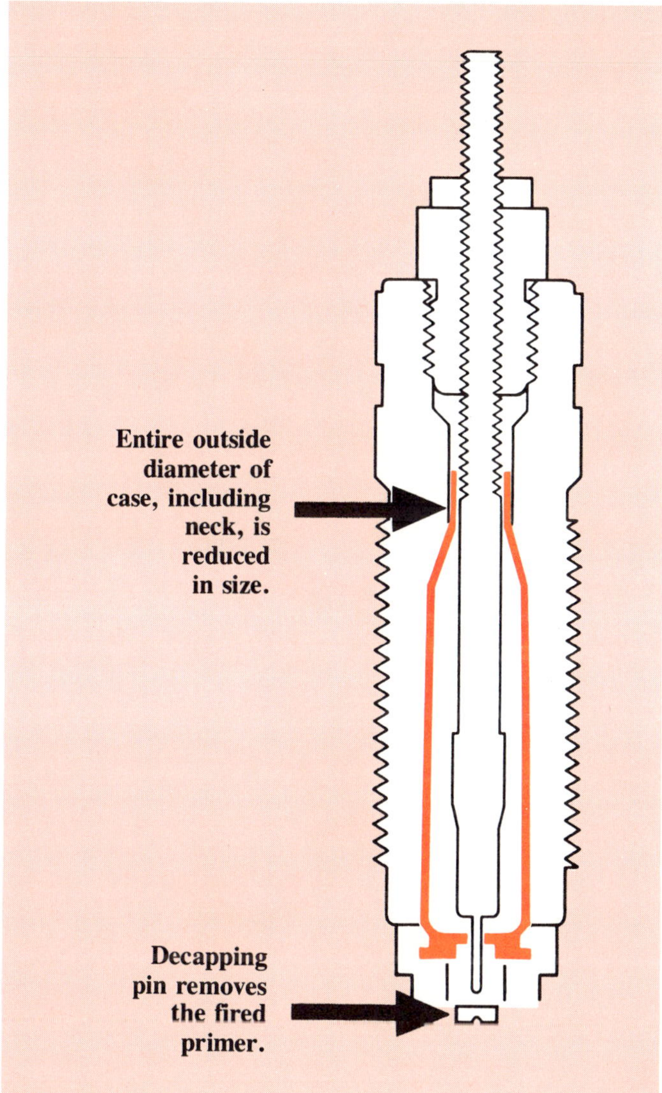

STEP 2

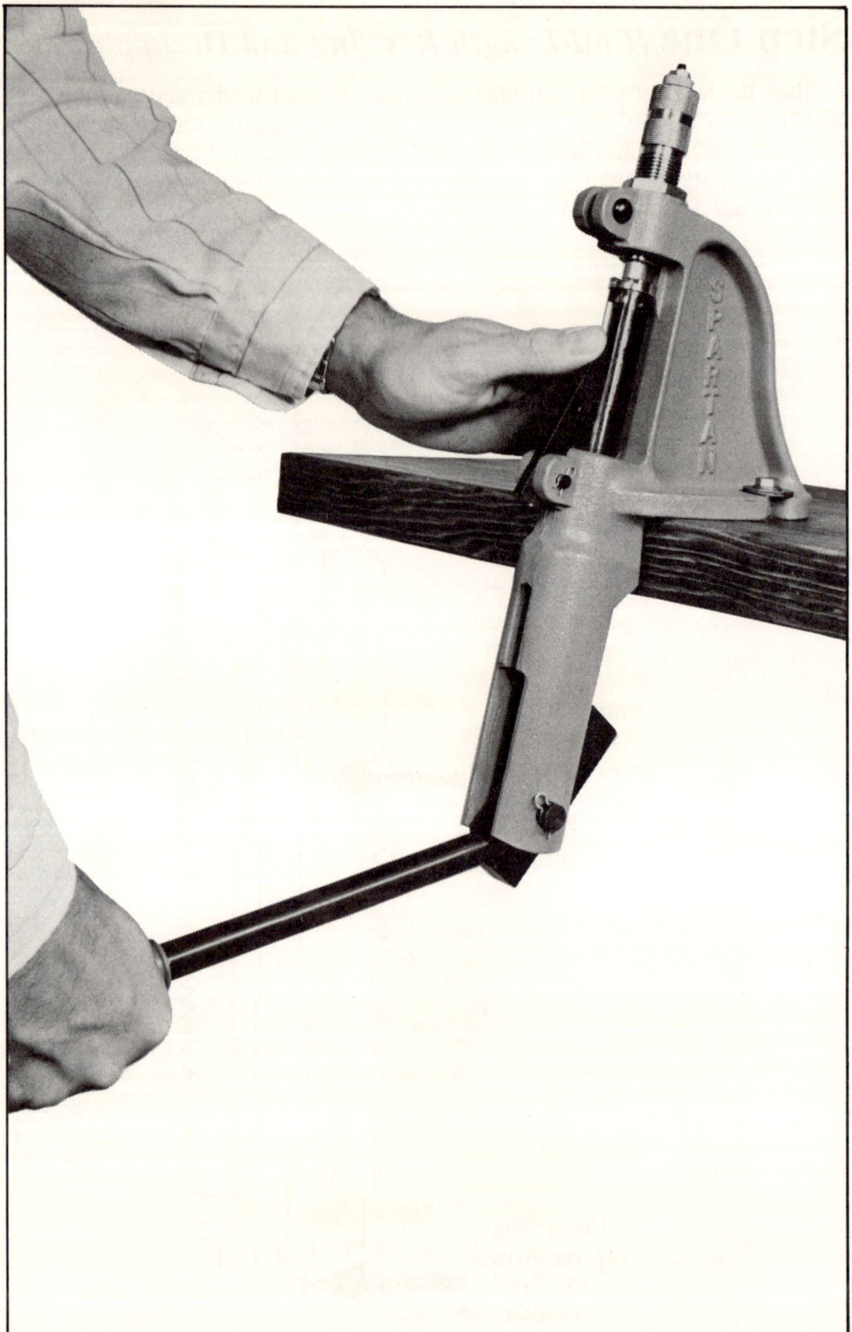

Step Two *(Inside Neck Expanding and Priming)*

As your case is withdrawn from the resizing die, two further operations are accomplished. The expanding button will automatically enlarge the neck, as shown in the cutaway drawing, and the priming punch will seat the new primer. As the expanding action of the button is automatic, you need not be concerned with it. You must, however, place the new primer (cup side up) into the priming punch sleeve. Push the priming arm forward (toward the press) and pull up on the press handle. As the ram is lowered, the priming arm will enter the slot in the side of the ram and seat the primer.

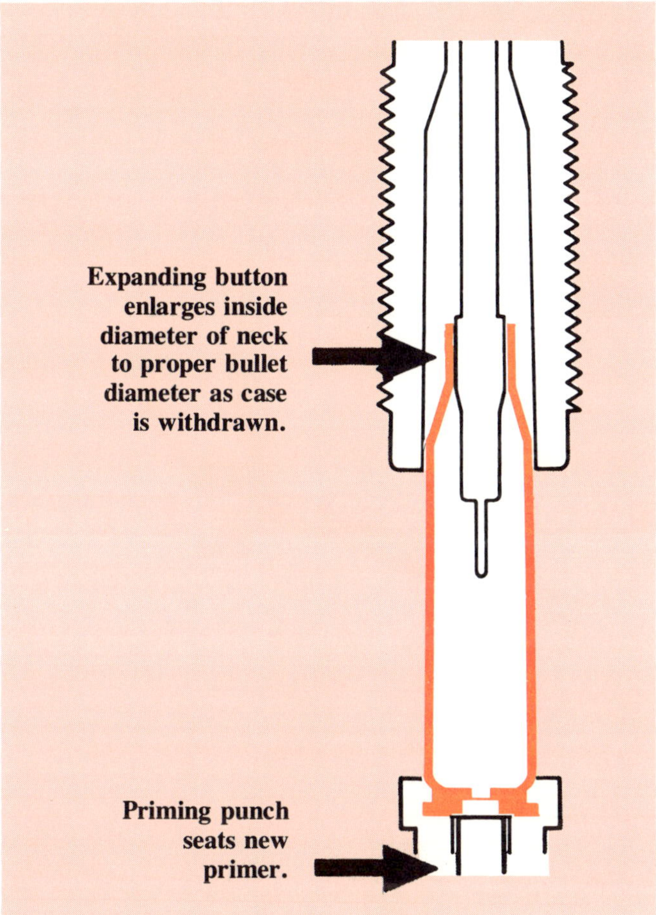

Seating Primers: Primers are seated mainly by feel. The bottom of the anvil **must bottom** in the primer pocket. Use care and do not crush the primer. Crushed primers give erratic ignition, or fail to fire.

Step Three *(Charging Powder)*

You will need an accurate powder scale such as the Lyman D-7. The data section specifies the powders appropriate for your particular cartridge. It also lists a suggested weight of the powder charge in grains and in fractions of grains. For example, 9.5 grains would read as NINE and FIVE TENTHS grains. 10.0 grains would be read as TEN grains. We recommend the novice restrict himself to the suggested starting load. Carefully level the powder scale as explained in the scale instructions and set it to weigh your required charge.
charge.

Slowly sprinkle small amounts of powder into the scale pan until the beam comes into balance. The beam is in balance when the pointed end (extreme left) is exactly on the zero mark.

Carefully remove the pan and pour its contents into the cartridge case. Use a powder funnel to make sure all the powder enters the case. To avoid the possibility of accidentally ''Double Charging'' a cartridge, you should develop a foolproof system of loading. A suggested method is to place all the uncharged cases on your left. As you pick up each case for charging, turn it up-side-down and shake it. This will insure that the case is empty. Turn the case right-side-up, charge it and place it carefully on your right. Take care when removing or replacing the scale pan that the poise are not accidentally moved.

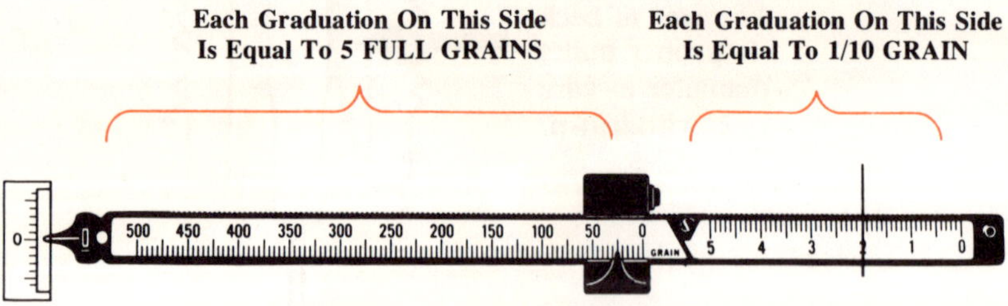

This illustration shows the beam of a modern reloading scale. Note how it is graduated on both sides of the pivot point. The scale is set by moving the two weights (poise) to the proper graduations. The large poise (on the left) is used to obtain multiples of FIVE GRAINS, while the small poise (on the right) is used for 1/10 FRACTIONS of a grain, or SINGLE grains from one to five.

EXAMPLE: The illustration shows a setting of 27.0 grains. If you wanted to decrease this 1/10 grain, you would simply move the small poise one notch to the right.

CAUTION: Technicians in the Lyman Lab have observed a potentially serious phenomena involving powder scales and plastic loading blocks and/or styrofoam packaging. It seems these substances sometimes retain a certain amount of static electricity, enough to create a magnetic field of varying radius.

This magnetic field has proven capable of radical deflection of uncharged and "zeroed" scales of all brands (available to us at the time). Of course, powder in the pan will tend to dampen the deflection but some still occurs depending on the charging level. Generally, the heavier the charge the less error . . . assuming the scale was first "zeroed" correctly.

We suggest you clear the surface of your reloading bench and make very sure the scale is set up accurately. Then move your equipment back piece by piece, paying particular attention to the plastic or styrofoam mentioned earlier. Please note this caution applies to all forms of reloading, not just metallic cartridges.

STEP 4

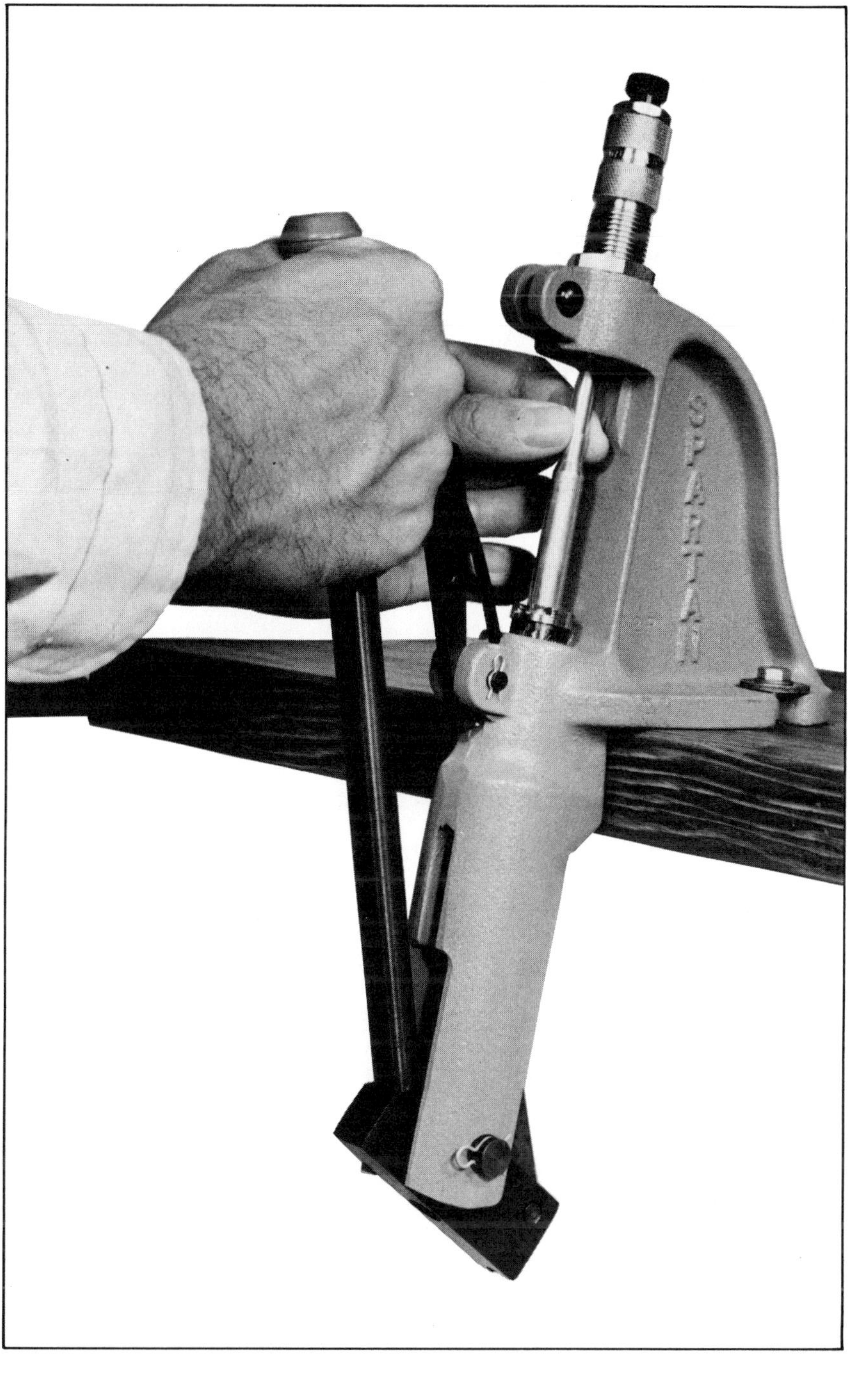

Step Four *(Bullet Seating)*

The last step in reloading a cartridge is seating the new bullet. Make certain that the overall length of the finished round is not longer than the maximum overall length listed in the data section.

This illustration shows how a bullet is seated. Screw the bullet seating die into the head of the press and adjust it according to the instructions supplied with the die. Place a primed, charged cartridge case in the shell holder and a bullet on the mouth of the case. Hold the bullet in place as you pull the press handle all the way down. As the case enters the die, the bullet will be pushed firmly into the neck of the case. Adjusting the seating screw controls the depth to which the bullet is seated. Adjusting the die body controls the crimp.

Crimping is a matter of choice and the seating die may be adjusted to crimp, or not to crimp as you desire. If you are loading hunting loads that will see hard usage in the magazine, it is wise to crimp-in the bullet. This prevents the bullets from unseating when the rifle is under recoil. Best accuracy, however, is usually obtained by not crimping-in the bullet. Target, or varmint loads, are best left uncrimped.

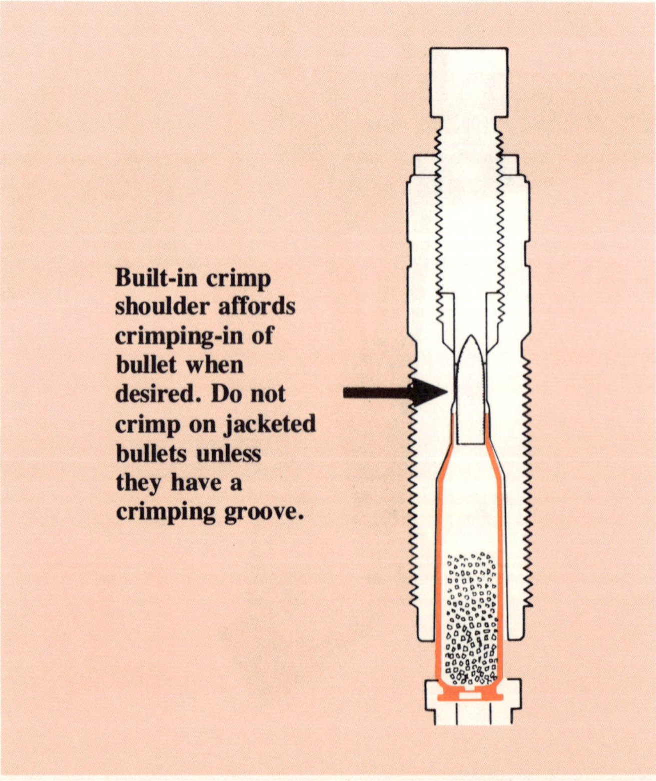

Reloading Data Introduction

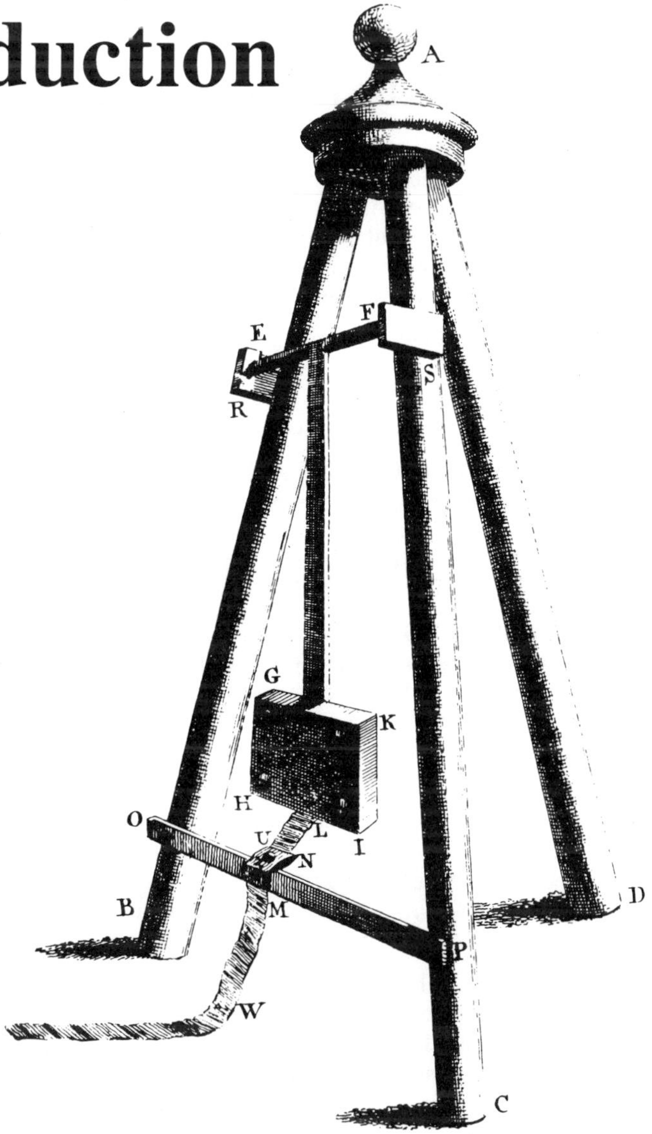

The Ballistic Pendulum as invented by Benjamin Robbins in 1742. This device allowed, for the first time, reasonably accurate measurement of projectile velocity.

DATA INTRODUCTION

The data listed in this section has been tested by our technicians and found to be safe when loaded with our test components and fired (under our laboratory controlled conditions) in our testing equipment. Since Lyman Products has no control over the manufacture of the various components listed, nor over the actual loading, choice or condition of the firearms and components used, no responsibility for use of this data is implied or assumed.

COMPONENTS: The components listed are not of Lyman manufacture. It is possible that production changes affecting ballistic performance can occur at any time without our knowledge. If there is ever a question as to the correctness of the component specified, write to its manufacturer.

STARTING LOAD: It is essential that the reader begin with the suggested weight of powder listed in this bracket and work up slowly to his best performing load. The novice should use only the ''starting load'' for a period of time until he builds confidence and experience. Never decrease this charge as an increase in pressure could be encountered.

MAXIMUM LOAD: All loads which are listed as maximum were tested and classified as maximum by our technicians in accordance with our laboratory standards. Under no circumstances should these loads be exceeded, nor should they be quickly accepted by the reader as a safe working maximum for his particular rifle or handgun.

Many reloaders misinterpret the meaning of the ''maximum load''. They wrongly assume that if a high pressure load proved safe in a test laboratory then it is equally safe under any and all conditions. This is not true. The reader must start with the ''starting load'' and work up his loads carefully. Working with his particular firearm and component combination, he may encounter signs of excess pressure before he reaches the maximum charge listed.

ACCURACY LOAD: As accuracy is not consistently the same with all firearms, this load requires some interpretation by the reader. Where an accuracy load is listed, it merely indicates our most accurate test results with this particular bullet weight and caliber.

Slight variations of the load may be necessary to produce optimum accuracy for your rifle or handgun. Where this load is printed in a colored panel, it indicates that the load was at, or near, a maximum pressure level. Such a load requires that the reader work up to it slowly.

VELOCITIES: The velocities are instrumental (not corrected to the muzzle) and were recorded at a distance of fifteen feet from the muzzle of the test firearm.

POWDERS: The following brands are listed: Dupont, Winchester, Hercules, Alcan, and Hodgdon.

BULLETS: The brand, weight and style of the various jacketed and cast bullets used in our tests are listed with the specifications under each cartridge heading.

COMPRESSED LOADS: All compressed loads are indicated with an asterisk. Depending upon the volume of the specific cartridge case used by the reader, he may, or may not, have difficulty starting bullets in such loads. If the bullet will not start, reduce the load sufficiently so that 1/10'' of space remains in the case neck. Start the bullet into the case and use whatever additional pressure is required to fully seat the bullet. Failure to comply could result in a bulged case.

.222 REMINGTON
(5.7 x 43mm)

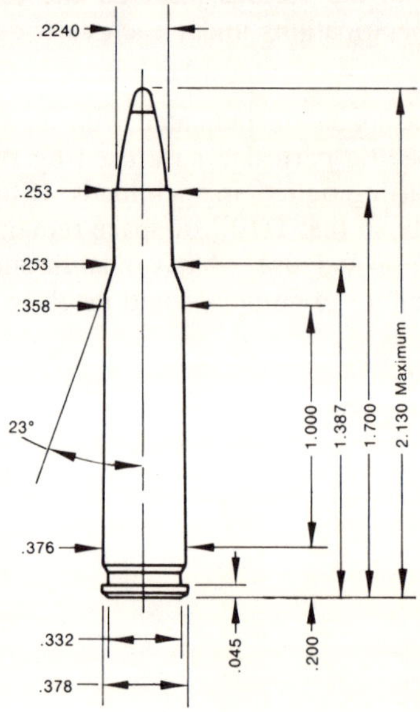

TEST COMPONENTS:

Cases .... Remington
Primers .... Remington 7½
Jacketed Bullets ....
Sierra 50 grain Spitzer;
Sierra 53 grain H.P.;
Sierra 55 grain Spitzer;
Hornady 60 grain Spire Pt.;

TEST (Pressure and Velocity) SPECIFICATIONS:

Firearm Used .... Universal Receiver
Barrel Length .... 26''
Twist .... 1-14''
Groove Diameter .... .224''

Note: All loads were tested with the Remington 7½ primer (designed specifically for cartridges of this type). I.M.R. 4198 is perhaps the most popular powder for this cartridge, however, all of the powders tested performed very well.

50 Grain Jacketed

Powder	Sug. Starting Grains	Velocity F.P.S.	Pressure C.U.P.	Max. Grains	Velocity F.P.S.	Pressure C.U.P.
IMR 4198	18.0	2717	28,200	20.5	3205	45,900
IMR 3031	20.0	2538	26,400	23.5*	3115	42,000
IMR 4064	20.0	2314	23,400	23.6*	2849	34,000
IMR 4895	22.0	2762	31,500	25.0*	3105	42,400
IMR 4320	23.0	2754	34,000	26.0*	3048	42,400
Rx 7	18.0	2680	25,200	21.5	3311	46,300
Ball C²	23.0	2865	34,000	26.3*	3257	46,700
H 335	23.0	2816	32,000	26.8*	3279	46,700

Accuracy Loads:
IMR 4198 Powder, 18.0 Grains, 2717 F.P.S., 28,200 C.U.P.
IMR 3031 Powder, 20.0 Grains, 2538 F.P.S., 26,400 C.U.P.
IMR 4895 Powder, 25.0* Grains, 3105 F.P.S., 42,400 C.U.P.

Factory Velocity Duplication Load:
IMR 4320 Powder, 26.0* Grains, 3048 F.P.S., 42,400 C.U.P.

Note: Loads shown in colored panels are maximum.
*Designates a compressed powder charge.

.222 REMINGTON Continued
(5.7 x 43mm)

53 Grain Jacketed

Powder	Sug. Starting Grains	Velocity F.P.S.	Pressure C.U.P.	Max. Grains	Velocity F.P.S.	Pressure C.U.P.
IMR 4198	18.0	2652	29,400	20.5	3125	46,300
IMR 3031	20.0	2544	27,600	23.5*	3039	41,100
IMR 4064	20.0	2341	25,800	23.6*	2840	36,900
IMR 4895	22.0	2747	34,000	25.0*	3058	45,900
IMR 4320	23.0	2672	35,000	26.0*	3039	46,700
Rx 7	18.0	2645	28,800	21.4	3205	46,300
Ball C²	22.0	2638	30,000	25.7	3154	46,700
H 335	22.0	2645	30,000	26.5*	3205	46,700

Accuracy Loads:
IMR 4895 Powder, 22.0 Grains, 2747 F.P.S., 34,000 C.U.P.
IMR 4198 Powder, 20.5 Grains, 3125 F.P.S., 46,300 C.U.P.

55 Grain Jacketed

Powder	Sug. Starting Grains	Velocity F.P.S.	Pressure C.U.P.	Max. Grains	Velocity F.P.S.	Pressure C.U.P.
IMR 4198	17.0	2538	27,000	20.2	3086	46,300
IMR 3031	20.0	2531	28,800	23.5*	3048	45,400
IMR 4064	20.0	2347	25,800	23.6*	2849	37,700
IMR 4895	21.0	2610	32,000	25.0*	3058	46,700
IMR 4320	22.5	2659	35,000	26.0*	3012	46,300
Rx 7	18.0	2717	30,000	21.0	3174	46,300
Ball C²	22.0	2680	32,000	25.0	3086	45,900
H 335	21.5	2544	29,400	26.3*	3115	46,300

Accuracy Loads:
IMR 4320 Powder, 22.5 Grains, 2659 F.P.S., 35,000 C.U.P.
IMR 4198 Powder, 20.2 Grains, 3086 F.P.S., 46,300 C.U.P.

60 Grain Jacketed

Powder	Sug. Starting Grains	Velocity F.P.S.	Pressure C.U.P.	Max. Grains	Velocity F.P.S.	Pressure C.U.P.
IMR 4198	16.5	2386	25,800	19.6	2906	46,300
IMR 3031	19.5	2433	29,400	23.5*	2994	46,700
IMR 4064	19.5	2267	28,200	23.6*	2785	41,600
IMR 4895	20.5	2469	32,000	24.1*	2906	46,700
IMR 4320	21.5	2450	34,000	25.1*	2873	46,300
Rx 7	17.0	2590	31,000	20.1	2985	46,300
Ball C²	20.0	2433	28,200	24.1	2949	46,700
H 335	21.0	2487	28,800	25.0	3012	45,900

Accuracy Loads:
IMR 4895 Powder, 20.5 Grains, 2469 F.P.S., 32,000 C.U.P.
Rx 7 Powder, 20.1 Grains, 2985 F.P.S., 46,300 C.U.P.

Note: Loads shown in colored panels are maximum.
*Designates a compressed powder charge.

.22/250

(.22 Varminter)

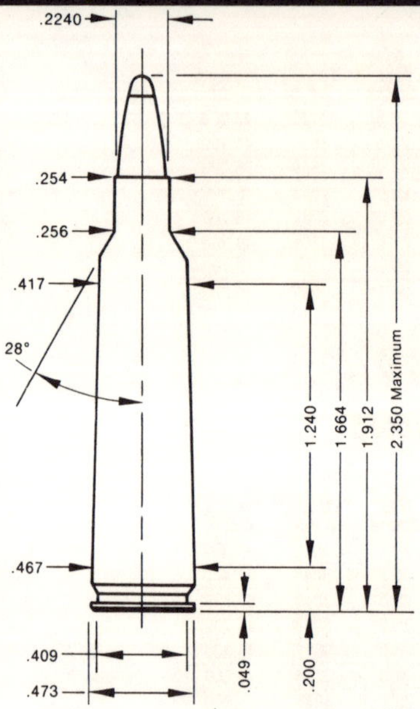

TEST COMPONENTS:

Cases .... Winchester and Remington
Primers ...... Winchester and Remington
Jacketed Bullets Used ....
Sierra 50 gr. Spitzer;
Sierra 55 gr. Spitzer;
Hornady 60 gr. Spire;
Sierra 63 gr. Semi-pointed

TEST (Pressure and Velocity) SPECIFICATIONS:

Firearm Used .... Universal Receiver
Barrel Length .... 26''
Twist .... 1-14''
Groove Diameter .... .224''

NOTE: With jacketed bullets best ballistic uniformity began with bullets weighing 53 grains. The lighter bullets shot well enough but gave wider velocity variations in a given string. All of our loads for this cartridge provided excellent accuracy.

Due to the wide variation in ''custom'' chambers, maximum loads in ''custom'' guns should be approached with extreme caution.

50 Grain Jacketed

Powder	Sug. Starting Grains	Velocity F.P.S.	Pressure C.U.P.	Max. Grains	Velocity F.P.S.	Pressure C.U.P.
IMR 3031	30.0	3181	33,000	35.3	3861	52,400
IMR 4895	31.0	3314	37,700	36.3	3832	53,600
IMR 4064	32.0	3210	32,000	37.0*	3854	52,400
IMR 4320	33.0	3408	40,700	37.0	3742	51,200
IMR 4350	36.0	3182	34,000	40.0*	3591	45,000
Rx 7	28.0	3478	41,100	31.5	3779	52,400
Ball C²	34.0	3480	36,900	38.5	3911	51,600
H 380	36.0	3569	39,400	40.0	3912	52,000
H 450	40.0	3363	39,400	45.5*	3763	52,400

Accuracy Loads:
IMR 4350 Powder, 40.0* Grains, 3591 F.P.S., 45,000 C.U.P.
H 380 Powder, 36.0 Grains, 3569 F.P.S., 39,400 C.U.P.

Note: Loads shown in colored panels are maximum.
*Designates a compressed powder charge.

.22/250 Continued
(.22 Varminter)

55 Grain Jacketed

Powder	Sug. Starting Grains	Velocity F.P.S.	Pressure C.U.P.	Max. Grains	Velocity F.P.S.	Pressure C.U.P.
IMR 3031	30.0	3217	36,400	34.3	3715	53,600
IMR 4895	31.0	3197	36,900	35.5	3693	53,200
IMR 4064	31.0	3161	35,000	36.0	3715	52,400
IMR 4320	32.0	3163	36,400	36.5	3630	52,000
IMR 4350	36.0	3195	37,700	40.0*	3561	49,600
Rx 7	26.5	3295	38,600	30.5	3615	53,600
Ball C²	34.0	3478	42,400	38.0	3802	53,600
H 380	35.0	3463	41,600	39.2	3799	53,600
H 450	39.0	3253	39,900	44.5*	3671	52,800

Accuracy Loads:
IMR 4064 Powder, 31.0 Grains, 3161 F.P.S., 35,000 C.U.P.
IMR 4895 Powder, 35.5 Grains, 3693 F.P.S., 53,200 C.U.P.
H 380 Powder, 39.2 Grains, 3799 F.P.S., 53,600 C.U.P.

60 Grain Jacketed

Powder	Sug. Starting Grains	Velocity F.P.S.	Pressure C.U.P.	Max. Grains	Velocity F.P.S.	Pressure C.U.P.
IMR 3031	30.0	3193	37,700	34.0	3596	52,800
IMR 4895	31.0	3173	38,100	35.0	3563	53,600
IMR 4064	31.0	3128	36,400	35.5	3589	52,000
IMR 4320	31.0	3064	37,300	36.0	3503	51,600
IMR 4350	36.0	3136	38,600	40.0*	3497	50,400
Rx 7	25.0	3080	39,000	29.0	3431	53,600
Ball C²	33.0	3346	45,000	36.7	3619	53,200
H 380	34.5	3362	44,200	38.0	3622	52,400
H 450	38.5	3188	40,300	43.5	3544	52,800

Accuracy Loads:
IMR 4350 Powder, 40.0* Grains, 3497 F.P.S., 50,400 C.U.P.
H 450 Powder, 38.5 Grains, 3188 F.P.S., 40,300 C.U.P.

63 Grain Jacketed

Powder	Sug. Starting Grains	Velocity F.P.S.	Pressure C.U.P.	Max. Grains	Velocity F.P.S.	Pressure C.U.P.
IMR 3031	29.0	3045	36,000	33.5	3507	53,600
IMR 4895	30.0	3056	36,900	34.5	3479	52,400
IMR 4064	31.0	3090	36,400	35.0	3490	50,800
IMR 4320	31.0	3011	36,900	35.8	3480	52,800
IMR 4350	36.0	3149	40,700	40.0*	3464	51,200
Rx 7	25.0	2986	38,600	28.5	3315	50,800
Ball C²	33.0	3223	42,400	36.0	3550	53,600
H 380	34.0	3268	43,700	37.5	3554	52,800
H 450	37.5	3070	38,100	43.0	3481	52,800

Accuracy Loads:
IMR 4895 Powder, 34.5 Grains, 3479 F.P.S., 52,400 C.U.P.
IMR 4350 Powder, 36.0 Grains, 3149 F.P.S., 40,700 C.U.P.
H 380 Powder, 37.5 Grains, 3554 F.P.S., 52,800 C.U.P.
H 450 Powder, 37.5 Grains, 3070 F.P.S., 38,100 C.U.P.

Note: Loads shown in colored panels are maximum.

*Designates a compressed powder charge.

.243 WINCHESTER

(6.2 x 52mm)

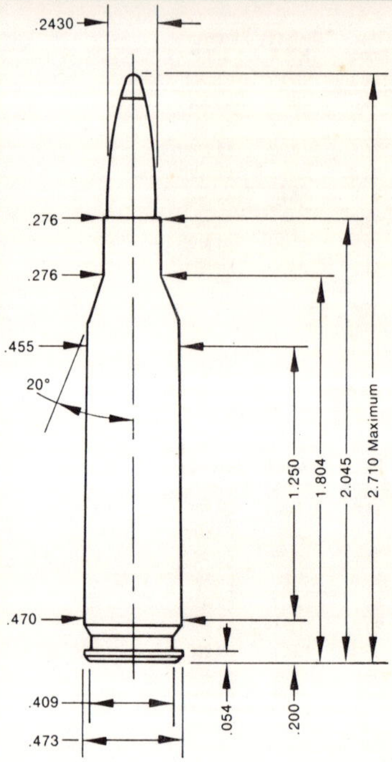

TEST COMPONENTS:

Cases .... Remington (all bullets except 100 gr. jacketed)
Winchester (100 gr. jacketed)

Primers .... Remington 9½

Jacketed Bullets Used .. Sierra 75 gr. H.P.; Sierra 85 gr. H.P.B.T.; Remington 90 gr. Ptd. S.P. No. B22722; Winchester 100 gr. Power Pt. S.P.

TEST (Pressure and Velocity) SPECIFICATIONS:

Firearm Used .... Universal Receiver

Barrel Length .... 26''

Twist .... 1-10''

Groove Diameter .... .243''

NOTE: If your .243 rifle is imported, we recommend you carefully check groove diameter. We have encountered rifles with groove diameters as small as .239. When fired in these tight groove rifles, factory loaded .243 cartridges blew primers and even our starting loads proved too hot. We, therefore, advise that you consider any rifle with a groove diameter of under .241 as dangerous.

In ''standard groove'' domestic rifles all of the I.M.R. powders gave excellent results. For long range varmint shooting, we recommend the use of 80 or 85 grain bullets as they do not lose their velocity as rapidly as the lighter projectiles.

.243 WINCHESTER Continued
(6.2 x 52mm)

75 Grain Jacketed

Powder	Sug. Starting Grains	Velocity F.P.S.	Pressure C.U.P.	Max. Grains	Velocity F.P.S.	Pressure C.U.P.
IMR 3031	34.0	3097	38,600	38.0	3407	49,000
IMR 4064	36.0	3127	39,000	40.5	3460	50,400
IMR 4320	37.0	3155	42,000	41.0	3437	50,000
IMR 4895	34.0	3041	37,700	39.0	3407	50,800
IMR 4350	40.0	3160	44,200	45.0	3425	51,200
Ball C²	34.0	2974	41,100	38.0	3264	50,400
H 380	36.0	3020	42,000	40.0	3294	52,000
H 450	45.0	3192	39,900	50.0	3539	52,400
4831	43.0	3099	41,100	48.0*	3486	51,600

Accuracy Loads:
IMR 4895 Powder, 39.0 Grains, 3407 F.P.S., 50,800 C.U.P.
H 380 Powder, 40.0 Grains, 3294 F.P.S., 52,000 C.U.P.
H 450 Powder, 45.0 Grains, 3192 F.P.S., 39,900 C.U.P.

85 Grain Jacketed

Powder	Sug. Starting Grains	Velocity F.P.S.	Pressure C.U.P.	Max. Grains	Velocity F.P.S.	Pressure C.U.P.
IMR 3031	32.0	2836	36,900	37.5	3190	49,200
IMR 4064	34.0	2824	33,500	39.0	3236	49,200
IMR 4320	35.0	2920	38,600	40.0	3255	50,000
IMR 4895	33.0	2883	38,600	38.0	3211	49,600
IMR 4350	38.0	2844	36,900	44.0	3283	51,200
Ball C²	31.0	2620	34,000	36.5	3035	49,200
H 380	33.0	2686	36,400	38.5	3094	50,400
H 450	43.0	3062	41,100	48.0	3374	52,800
4831	41.0	2832	36,400	46.5*	3266	49,600

Accuracy Loads:
IMR 4350 Powder, 38.0 Grains, 2844 F.P.S., 36,900 C.U.P.

90 Grain Jacketed

Powder	Sug. Starting Grains	Velocity F.P.S.	Pressure C.U.P.	Max. Grains	Velocity F.P.S.	Pressure C.U.P.
IMR 3031	32.0	2778	37,700	37.0	3084	49,200
IMR 4064	35.0	2899	40,300	38.5	3097	49,200
IMR 4320	35.0	2856	38,600	39.5	3146	49,200
IMR 4895	33.0	2758	38,100	37.5	3106	49,200
IMR 4350	37.0	2703	34,000	43.5	3161	49,200
Ball C²	31.0	2560	34,000	36.0	2902	48,800
H 380	33.0	2598	33,500	38.0	2922	48,800
H 450	42.0	2899	38,100	47.5	3222	48,000
4831	41.0	2734	33,500	46.0*	3080	49,600

Accuracy Loads:
IMR 4320 Powder, 35.0 Grains, 2856 F.P.S., 38,600 C.U.P.
IMR 4350 Powder, 42.5 Grains, 3090 F.P.S., 46,700 C.U.P.

Note: Loads shown in colored panels are maximum.
*Designates a compressed powder charge.

.243 WINCHESTER Continued
(6.2 x 52mm)

100 Grain Jacketed

Powder	Sug. Starting Grains	Velocity F.P.S.	Pressure C.U.P.	Max. Grains	Velocity F.P.S.	Pressure C.U.P.
IMR 3031	30.5	2594	42,000	35.0	2869	50,800
IMR 4064	32.0	2658	43,300	36.5	2912	52,400
IMR 4320	34.5	2694	43,700	39.0	3004	52,800
IMR 4895	31.5	2604	39,400	36.5	2919	52,800
IMR 4350	37.0	2668	40,300	42.5	3011	52,800
Ball C²	30.5	2463	40,700	34.5	2723	50,800
H 380	32.0	2549	41,600	36.0	2781	52,400
H 450	40.0	2739	40,300	46.5	3065	52,800
4831	40.0	2684	39,900	45.0	3010	51,600

Accuracy Loads:
IMR 4320 Powder, 34.5 Grains, 2694 F.P.S., 43,700 C.U.P.
IMR 4350 Powder, 40.5 Grains, 2913 F.P.S., 50,000 C.U.P.
H 450 Powder, 46.5 Grains, 3065 F.P.S., 52,800 C.U.P.

Factory Velocity Duplication Load:
H 450 Powder, 41.5 Grains, 2826 F.P.S., 43,300 C.U.P.

Note: Loads shown in colored panels are maximum.

*Designates a compressed powder charge.

.270 WINCHESTER
(6.9 x 64mm)

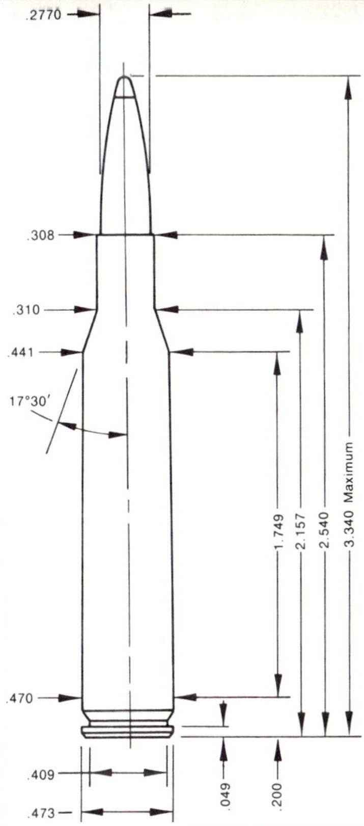

TEST COMPONENTS:

Cases ...Remington (Jacketed 100 to 170 gr.)
Winchester (Cast and 90 gr. Jacketed)
Primers .... Winchester 120
Jacketed Bullets Used ....
Winchester 100 gr. Ptd. S.P.;
Sierra 130 gr. Spitzer;
Winchester 150 gr. Power Point, S.P.;
Speer 170 gr. R.N.

TEST (Pressure and Velocity) SPECIFICATIONS:

Firearm Used .... Universal Receiver
Barrel Length .... 26''
Twist .... 1-10''
Groove Diameter .... .277''

NOTE: Jacketed bullets of 90 and 100 grains did not deliver as uniform ballistics or as good accuracy as the heavier bullets.

100 Grain Jacketed

Powder	Sug. Starting Grains	Velocity F.P.S.	Pressure C.U.P.	Max. Grains	Velocity F.P.S.	Pressure C.U.P.
IMR 3031	45.0	3154	43,700	50.0	3427	52,800
IMR 4895	46.5	3134	44,600	50.0	3356	53,600
IMR 4064	46.0	3067	40,300	51.5	3413	53,600
IMR 4320	46.0	2976	39,900	51.5	3333	53,200
IMR 4350	50.0	2949	41,100	57.0*	3345	54,400
H 380	48.0	3086	39,900	54.0	3436	54,000
H 450	53.0	2898	37,300	61.5	3413	54,400
4831	55.0	2702	40,300	62.0*	3356	51,600

Factory Velocity Duplication Load:
IMR 4320 Powder, 51.2 Grains, 3279 F.P.S., 51,600 C.U.P.

Note: Loads shown in colored panels are maximum.

*Designates a compressed powder charge.

.270 WINCHESTER
(6.9 x 64mm)

130 Grain Jacketed

Powder	Sug. Starting Grains	Velocity F.P.S.	Pressure C.U.P.	Max. Grains	Velocity F.P.S.	Pressure C.U.P.
IMR 3031	42.0	2739	41,600	47.0	3030	53,200
IMR 4895	42.0	2680	41,100	48.5	3030	54,400
IMR 4064	43.5	2724	39,000	49.0	3039	52,400
IMR 4320	44.0	2695	39,400	50.5	3067	54,400
IMR 4350	49.0	2739	39,400	55.5*	3134	54,400
H 380	44.0	2680	41,100	50.0	2994	54,000
H 450	51.0	2652	37,700	58.5	3105	54,000
4831	53.0	2762	40,700	60.0*	3125	53,600

Accuracy Load:
IMR 4064 Powder, 43.5 Grains, 2724 F.P.S., 39,000 C.U.P.

Factory Velocity Duplication Load:
IMR 4350 Powder, 55.5* Grains, 3134 F.P.S., 54,400 C.U.P.

150 Grain Jacketed

Powder	Sug. Starting Grains	Velocity F.P.S.	Pressure C.U.P.	Max. Grains	Velocity F.P.S.	Pressure C.U.P.
IMR 3031	38.0	2469	39,900	45.0	2801	54,400
IMR 4895	39.0	2457	39,400	46.0	2801	54,000
IMR 4064	42.0	2564	42,000	47.0	2824	52,800
IMR 4320	43.0	2583	42,400	48.5	2857	54,400
IMR 4350	47.0	2610	40,700	52.5	2906	54,000
H 380	40.0	2427	40,300	46.5	2739	54,400
H 450	49.5	2538	37,700	57.0	2932	53,200
4831	50.0	2557	39,000	57.0*	2941	54,000

Accuracy Load:
IMR 4064 Powder, 42.0 Grains, 2564 F.P.S., 42,000 C.U.P.

Factory Velocity Duplication Load:
IMR 4350 Powder, 52.0 Grains, 2842 F.P.S., 52,800 C.U.P.

170 Grain Jacketed

Powder	Sug. Starting Grains	Velocity F.P.S.	Pressure C.U.P.	Max. Grains	Velocity F.P.S.	Pressure C.U.P.
IMR 3031	37.5	2293	38,100	44.5	2631	53,600
IMR 4895	38.0	2277	39,000	45.5	2631	52,800
IMR 4064	40.0	2331	38,600	46.5	2659	52,000
IMR 4320	41.0	2358	37,300	48.0	2695	54,000
IMR 4350	46.0	2433	39,400	52.0	2739	52,000
H 380	39.0	2252	39,000	46.0	2564	54,400
H 450	49.0	2409	38,100	56.5	2777	52,800
4831	49.0	2398	36,000	56.0*	2801	53,600

Accuracy Load:
IMR 4350 Powder, 52.0 Grains, 2739 F.P.S., 52,000 C.U.P.

Note: Loads shown in colored panels are maximum.

*Designates a compressed powder charge.

.30/30 WINCHESTER
(7.62 x 51R mm)

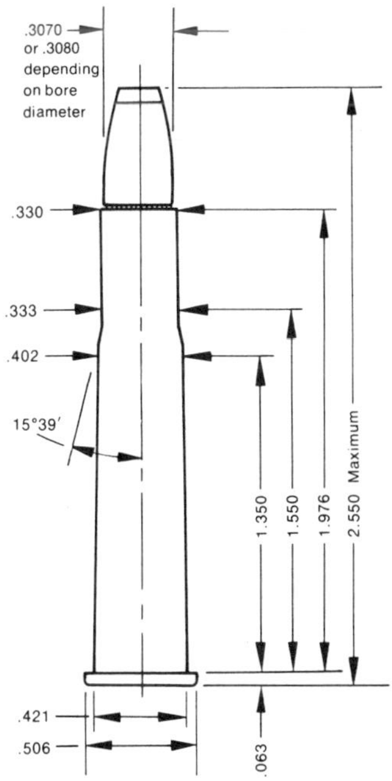

TEST COMPONENTS:

Cases . . . . Winchester
Primers . . . . Winchester 8½-120
Jacketed Bullets Used . . . .
Sierra 110 gr. R.N. (30 M1 Carbine);
Remington 150 gr. S.P.C.-L., No. B22774;
Winchester 170 gr. Power Point S.P.;

TEST (Pressure and Veolcity) SPECIFICATIONS:

Firearm Used . . . . Universal Receiver
Barrel Length . . . . 26''
Twist . . . . 1-12''
Groove Diameter . . . . .308''

NOTE: It was very difficult to obtain accuracy with the 110 gr. jacketed bullet. The 110 gr. jacketed bullet is the only bullet in this caliber which was not crimped in place for our test. All other bullets were crimped. Jacketed accuracy loads gave extreme velocity spreads from 4 F.P.S. to 25 F.P.S.

Accuracy with the 150 grain, 170 grain, and 190 grain bullet was superb.

All the IMR Series powders gave excellent accuracy with the 170 grain bullet.

The use of only flat or blunt nosed bullets is required in tubular magazines. Crimping is necessary to keep the bullet in place in tubular magazines.

Some older rifles for this cartridge are marked 30 W.C.F.

NOTE: Loads shown in colored panels are maximum.

*Designates a compressed powder charge.

.30/30 WINCHESTER Continued
(7.62 x 51R mm)

110 Grain Jacketed

Powder	Sug. Starting Grains	Velocity F.P.S.	Pressure C.U.P.	Max. Grains	Velocity F.P.S.	Pressure C.U.P.
IMR 4198	27.0	2358	28,200	29.5	2617	38,100
IMR 3031	31.0	2442	35,500	33.5	2623	38,900
IMR 4064	31.0	2168	27,000	35.0*	2568	37,300
IMR 4320	32.0	2281	32,000	35.0	2467	38,900
IMR 4895	31.0	2258	28,200	34.0	2545	37,700
IMR 4350	35.0	2181	29,400	39.0*	2419	35,000
Rx 7	27.5	2458	30,000	31.5	2801	38,600
Ball C²	34.5	2504	32,500	38.0	2722	38,900
H 450	37.0	2184	33,500	41.0*	2395	38,900

150 Grain Jacketed

Powder	Sug. Starting Grains	Velocity F.P.S.	Pressure C.U.P.	Max. Grains	Velocity F.P.S.	Pressure C.U.P.
IMR 4198	23.0	1991	29,400	26.5	2278	39,000
IMR 3031	26.5	1876	25,200	29.5	2274	39,400
IMR 4064	29.0	2055	32,000	32.0	2313	39,400
IMR 4320	29.0	1956	31,000	32.0	2218	39,400
IMR 4895	28.0	1980	29,400	31.0	2248	38,600
IMR 4350	32.0	1980	31,000	36.0*	2218	38,100
Rx 7	25.0	2154	30,500	28.0	2350	38,100
Ball C²	28.0	1989	29,400	31.5	2268	39,000
H 380	30.0	2093	32,500	33.0	2227	37,300
H 450	35.0	1982	35,000	38.0	2103	38,100

Accuracy Loads:
IMR 3031 Powder, 29.5 Grains, 2274 F.P.S., 39,400 C.U.P.
Ball C² Powder, 31.5 Grains, 2268 F.P.S., 39,000 C.U.P.

170 Grain Jacketed

Powder	Sug. Starting Grains	Velocity F.P.S.	Pressure C.U.P.	Max. Grains	Velocity F.P.S.	Pressure C.U.P.
IMR 3031	25.0	1786	27,600	28.5	2110	38,600
IMR 4064	27.0	1845	28,200	30.5	2150	38,100
IMR 4320	27.0	1771	28,200	30.5	2050	38,600
IMR 4895	26.0	1837	30,000	29.5	2091	39,000
IMR 4350	30.0	1816	28,200	34.5*	2104	38,100
Rx 7	23.0	1961	32,500	26.0	2127	37,700
Ball C²	27.0	1869	27,000	30.5	2128	37,300
H 380	28.0	1864	29,400	31.5	2101	38,100
H 450	32.0	1779	30,500	36.0	1992	39,400

.30/30 WINCHESTER Continued

(7.62 x 51R mm)

176 Grain Cast (w/Gas Check)
BULLET #31141 (#2 ALLOY)

Powder	Sug. Starting Grains	Velocity F.P.S.	Pressure C.U.P.	Max. Grains	Velocity F.P.S.	Pressure C.U.P.
Hi-Skor 700X	6.0	1124	24,600	7.5	1269	34,000
PB	6.5	1100	23,400	8.7	1286	36,900
SR 7625	6.5	1118	22,800	9.0	1337	36,900
SR 4756	7.0	1102	19,800	10.5	1423	36,400
Red Dot	6.5	1159	24,600	8.5	1348	34,500
Unique	7.0	1240	21,600	10.6	1555	35,500
Green Dot	7.0	1192	25,200	9.0	1373	33,500
Herco	8.5	1281	25,800	11.0	1505	33,000

Accuracy Load:
Unique Powder, 7.0 Grains, 1240 F.P.S., 21,600 C.U.P.

NOTE: Loads shown in colored panels are maximum.

*Designates a compressed powder charge.

.308 WINCHESTER
(7.62 x 51mm NATO)

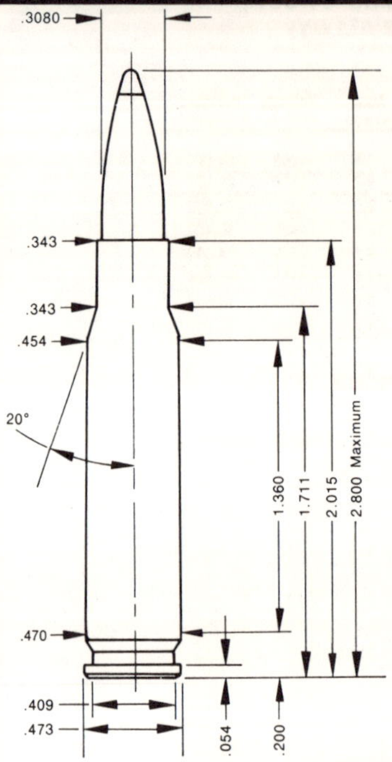

TEST COMPONENTS:

Cases ..... Winchester (Jacketed 110 gr. to 150 gr. and all cast bullets)
Remington (Jacketed 165 gr. to 220 gr.)
Primers ..... Winchester 8½-120 (Jacketed 110 gr. to 150 gr. and all cast bullets)
Remington 9½ (Jacketed 165 gr. to 220 gr.)
Jacketed Bullets Used ....
Remington 150 gr. Ptd. S.P.C.-L., No. B22776;
Hornady 165 gr. Spire Pt.;
Sierra 180 gr. Spitzer;
Sierra 190 gr. Match King H.P.;

TEST (Pressure and Velocity) SPECIFICATIONS:

Firearm Used ....Universal Receiver
Barrel Length .... 26''
Twist .... 1-12''
Groove Diameter .... .308''

NOTE: Accuracy loads with jacketed bullets showed velocity spreads of 6 F.P.S. to 23 F.P.S.

If military cases are used, maximum loads should be reduced by two (2) full grains.

This cartridge is capable of exceptionally fine accuracy with most component combinations.

150 Grain Jacketed

Powder	Sug. Starting Grains	Velocity F.P.S.	Pressure C.U.P.	Max. Grains	Velocity F.P.S.	Pressure C.U.P.
IMR 3031	40.0	2544	36,000	44.5	2849	52,800
IMR 4895	40.0	2403	32,000	45.0	2777	50,000
IMR 4064	43.0	2645	41,100	48.0*	2890	52,000
IMR 4320	43.0	2493	37,300	48.5*	2840	52,800
IMR 4350	46.0	2380	33,500	51.6*	2659	41,600
Rx 7	35.0	2475	35,000	39.0	2739	48,800
Ball C^2	45.0	2717	40,700	49.0	2915	52,400
H 335	46.0	2754	40,700	50.5	2985	52,000
H 380	47.0	2695	42,900	53.0*	2949	51,200
H 450	52.0	2512	40,700	58.5*	2777	50,800

Accuracy Loads:
IMR 4895 Powder, 45.0 Grains, 2777 F.P.S., 50,000 C.U.P.
Ball C^2 Powder, 49.0 Grains, 2915 F.P.S., 52,400 C.U.P.
H 380 Powder, 47.0 Grains, 2695 F.P.S., 42,900 C.U.P.
IMR 3031 Powder, 44.0 Grains, 2800 F.P.S., 50,000 C.U.P.

Factory Velocity Duplication Loads:
IMR 4064 Powder, 48.0* Grains, 2890 F.P.S., 52,000 C.U.P.

NOTE: Loads shown in colored panels are maximum.
*Designates a compressed powder charge.

.308 WINCHESTER Continued
(7.62 x 51mm NATO)

165 Grain Jacketed

Powder	Sug. Starting Grains	Velocity F.P.S.	Pressure C.U.P.	Max. Grains	Velocity F.P.S.	Pressure C.U.P.
IMR 3031	37.0	2331	35,000	42.0	2652	49,600
IMR 4895	38.0	2288	33,000	43.0	2645	52,800
IMR 4064	40.0	2421	36,900	45.5*	2732	52,800
IMR 4320	42.0	2463	42,400	45.0	2632	51,200
IMR 4350	46.0	2409	38,100	50.0*	2631	45,900
Rx 7	31.0	2277	37,300	36.5	2577	52,400
Ball C²	41.0	2398	35,000	46.0	2732	52,000
H 335	43.0	2544	39,900	47.5	2793	51,200
H 380	45.0	2551	41,700	51.0*	2793	50,000
H 450	50.0	2341	39,400	56.5*	2652	52,800

Accuracy Loads:
H 380 Powder, 51.0* Grains, 2793 F.P.S., 50,000 C.U.P.

180 Grain Jacketed

Powder	Sug. Starting Grains	Velocity F.P.S.	Pressure C.U.P.	Max. Grains	Velocity F.P.S.	Pressure C.U.P.
IMR 3031	35.0	2183	33,500	40.5	2538	51,200
IMR 4895	38.0	2267	36,000	42.5	2564	52,400
IMR 4064	40.0	2392	40,700	44.5*	2604	51,600
IMR 4320	40.0	2293	38,600	44.5	2557	52,400
IMR 4350	45.0	2277	36,000	51.0*	2624	51,600
Ball C²	39.0	2293	36,900	44.5	2583	51,200
H 335	40.0	2304	35,500	45.0	2631	52,000
H 380	42.0	2392	39,900	48.5*	2652	52,800
H 450	46.0	2227	38,600	53.0*	2493	50,800
4831	45.0	2096	33,000	50.5*	2398	43,300

Accuracy Loads:
H 380 Powder, 42.0 Grains, 2392 F.P.S., 39,900 C.U.P.
IMR 4895 Powder, 41.0 Grains, 2481 F.P.S., 44,200 C.U.P.

Factory Velocity Duplication Loads:
IMR 4064 Powder, 44.5* Grains, 2604 F.P.S., 51,600 C.U.P.
H 380 Powder, 48.0* Grains, 2604 F.P.S., 48,800 C.U.P.

190 Grain Jacketed

Powder	Sug. Starting Grains	Velocity F.P.S.	Pressure C.U.P.	Max. Grains	Velocity F.P.S.	Pressure C.U.P.
IMR 3031	35.0	2150	33,000	39.5	2469	50,000
IMR 4895	37.0	2217	35,500	41.5	2500	51,200
IMR 4064	39.0	2331	40,300	43.0*	2544	52,800
IMR 4320	39.0	2247	39,000	43.0	2481	50,800
IMR 4350	43.0	2227	40,700	48.5*	2518	52,800
Ball C²	38.0	2207	35,000	43.5	2557	51,200
H 335	39.0	2242	35,000	44.0	2564	51,600
H 380	43.0	2409	42,400	48.0*	2624	51,200
H 450	46.0	2178	38,100	53.0*	2487	52,000
4831	45.0	2127	33,500	50.0*	2398	43,300

Accuracy Loads:
IMR 4895 Powder, 41.5 Grains, 2500 F.P.S., 51,200 C.U.P.
H 380 Powder, 43.0 Grains, 2409 F.P.S., 42,400 C.U.P.

NOTE: Loads shown in colored panels are maximum.
*Designates a compressed powder charge.

.30/06
(7.62 x 63mm)

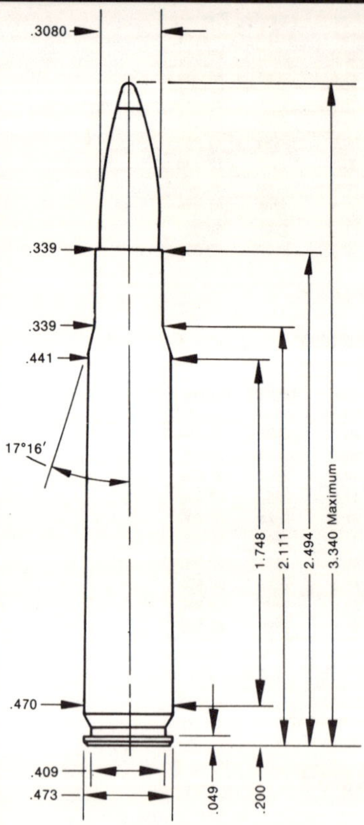

TEST COMPONENTS:

Cases .... Winchester (all tests except 220 gr. Jacketed)
Remington (220 gr. Jacketed)
Primers .... Winchester 8½-120
Jacketed Bullets Used ....
Hornady 165 gr. Spire Point;
Sierra 180 gr. Spitzer;
Sierra 190 gr. Match King H.P.;
Winchester 220 gr. Power Point S.P.

TEST (Pressure and Velocity) SPECIFICATIONS:

Firearm Used .... Universal Receiver
Barrel Length .... 26''
Twist .... 1-10''
Groove Diameter .... .308''

NOTE: Jacketed bullet accuracy loads were held to 40 F.P.S. spread or less, some spreads being as little as 8 F.P.S.

Ballistically, best uniformity with the 110 grain jacketed bullet was obtained with IMR 4895. Best results with the 125 and 130 grain jacketed bullets were with Reloader 11. For the 150 grain bullet IMR 4064 and 4350 proved excellent. The entire IMR series along with Reloader 11, 760 BR and H380 worked well with the 168 grain projectile. Of the bullets tested, the 168 grain Sierra International was, by far, the most accurate in this caliber.

.30/06 Continued
(7.62 x 63mm)

165 Grain Jacketed

Powder	Sug. Starting Grains	Velocity F.P.S.	Pressure C.U.P.	Max. Grains	Velocity F.P.S.	Pressure C.U.P.
IMR 3031	42.0	2481	35,500	48.0	2801	48,000
IMR 4895	43.0	2469	36,400	48.5	2732	46,000
IMR 4064	45.0	2531	34,500	50.0	2824	48,000
IMR 4320	45.0	2506	38,100	51.5	2832	50,800
IMR 4350	52.0	2583	38,100	57.0*	2824	48,000
Ball C²	43.0	2444	34,000	48.5	2754	48,000
H 380	45.0	2475	36,000	51.5	2824	50,000
4831	55.0	2557	35,500	61.5*	2881	45,900
H 450	53.0	2518	36,000	59.5	2840	50,800

Accuracy Loads:
IMR 4064 Powder, 48.5 Grains, 2732 F.P.S., 46,000 C.U.P.
IMR 4350 Powder, 52.0 Grains, 2583 F.P.S., 38,100 C.U.P.

168 Grain Jacketed

Powder	Sug. Starting Grains	Velocity F.P.S.	Pressure C.U.P.	Max. Grains	Velocity F.P.S.	Pressure C.U.P.
IMR 3031	42.0	2463	34,000	48.0	2808	49,600
IMR 4895	43.0	2444	35,500	48.5	2762	47,600
IMR 4064	45.0	2551	34,000	50.0	2816	46,700
IMR 4320	45.0	2475	36,400	51.0	2816	50,000
IMR 4350	52.0	2590	38,000	57.0*	2873	50,400
Ball C²	43.0	2538	42,000	48.5	2777	48,400
H 380	45.0	2512	37,700	51.5	2840	50,800
4831	55.0	2564	37,300	61.5*	2873	47,200
H 450	53.0	2518	35,000	59.5	2840	50,400

Accuracy Loads:
H 380 Powder, 51.5 Grains, 2840 F.P.S., 50,800 C.U.P.
IMR 4064 Powder, 45.0 Grains, 2551 F.P.S., 34,000 C.U.P.

190 Grain Jacketed

Powder	Sug. Starting Grains	Velocity F.P.S.	Pressure C.U.P.	Max. Grains	Velocity F.P.S.	Pressure C.U.P.
IMR 3031	40.0	2309	34,500	45.0	2590	47,000
IMR 4895	41.0	2288	35,000	46.5	2624	50,400
IMR 4064	42.0	2309	32,500	47.5	2624	48,400
IMR 4320	44.0	2341	34,500	49.0	2631	48,800
IMR 4350	50.0	2475	39,900	55.0*	2732	50,400
Ball C²	41.0	2247	34,000	46.5	2518	47,000
H 380	42.0	2309	37,300	47.5	2577	48,000
4831	53.0	2450	36,900	59.0*	2724	47,200
H 450	50.0	2347	36,000	55.5	2624	48,000

Accuracy Loads:
IMR 4320 Powder, 48.0 Grains, 2570 F.P.S., 45,000 C.U.P.
IMR 4350 Powder, 55.0* Grains, 2732 F.P.S., 50,400 C.U.P.
4831 Powder, 53.0 Grains, 2450 F.P.S., 36,900 C.U.P.

.30/06 Continued
(7.62 x 63mm)

220 Grain Jacketed

Powder	Sug. Starting Grains	Velocity F.P.S.	Pressure C.U.P.	Max. Grains	Velocity F.P.S.	Pressure C.U.P.
IMR 3031	38.0	2087	35,000	43.0	2336	48,000
IMR 4895	40.0	2197	41,100	44.5	2375	48,400
IMR 4064	41.0	2183	37,300	46.5	2439	49,200
IMR 4320	41.0	2118	36,400	47.5	2421	49,200
IMR 4350	49.0	2309	40,700	53.0*	2500	49,200
H 380	41.0	2145	39,400	45.5	2341	50,000
4831	51.0	2283	38,600	57.5*	2583	50,000
H 450	49.0	2327	37,300	54.5	2463	47,600

Accuracy Loads:
IMR 4320 Powder, 46.5 Grains, 2392 F.P.S., 45,400 C.U.P.
IMR 4350 Powder, 53.0* Grains, 2500 F.P.S., 49,200 C.U.P.

Factory Velocity Duplication Loads:
IMR 4350 Powder, 50.5 Grains, 2409 F.P.S., 44,200 C.U.P.
4831 Powder, 53.6 Grains, 2409 F.P.S., 48,800 C.U.P.

NOTE: Loads shown in colored panels are maximum.
*Designates a compressed powder charge.

9mm LUGER
(9mm Parabellum)

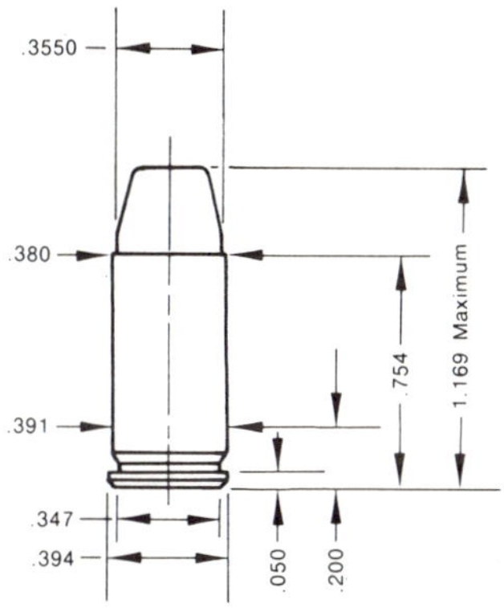

TEST COMPONENTS:

Cases .... Remington
Primers .... Remington 1½
Jacketed Bullets Used ....
Remington 95 grain M.C. No. B22844;
Remington 124 grain M.C. No. B22842

TEST (Velocity) SPECIFICATIONS:

Firearm Used . Smith and Wesson Model 39
Barrel Length .... 4''
Twist .... 1-10''
Groove Diameter .... .3535''

NOTE: Variations in groove diameters are often encountered with pistols chambered for this cartridge. However, the most common diameter is .354''. Jacketed bullets of .355'' may be used safely in all barrels ranging from .354'' to .356''. This particular cartridge headspaces on the case mouth and, therefore, the case should not be crimped. Also, the trim-to length should be adhered to closely. A short case, or a crimp on the mouth, can cause problems with headspace.

Our most popular cast bullet for this caliber is No. 356402.

95 Grain Jacketed

Powder	Sug. Starting Grains	Velocity F.P.S.	Max. Grains	Velocity F.P.S.
Bullseye	3.7	975	5.2	1273
Red Dot	3.8	926	5.3	1234
Unique	4.8	1000	6.2	1275

Accuracy Load:
Unique Powder, 6.2 Grains, 1275 F.P.S.

9mm LUGER Continued
(9mm Parabellum)

124 Grain Jacketed

Powder	Sug. Starting Grains	Velocity F.P.S.	Max. Grains	Velocity F.P.S.
Bullseye	3.5	909	4.8	1142
Red Dot	3.5	877	4.9	1107
Unique	4.3	891	5.8	1149

Accuracy Load:
Unique Powder, 5.5 Grains, 1096 F.P.S.

Factory Duplication Load:
Unique Powder, 5.5 Grains, 1096 F.P.S.

121 Grain Cast
BULLET #356402 (#2 ALLOY)
CAN ALSO USE BULLET #358242

Powder	Sug. Starting Grains	Velocity F.P.S.	Max. Grains	Velocity F.P.S.
Bullseye	3.5	990	4.8	1212
Red Dot	3.6	966	5.0	1197
Unique	4.4	999	6.0	1243
Herco	4.6	952	6.1	1183

Accuracy Load:
Bullseye Powder, 3.5 Grains, 990 F.P.S.

158 Grain Cast
BULLET #358311 (#2 ALLOY)

Powder	Sug. Starting Grains	Velocity F.P.S.	Max. Grains	Velocity F.P.S.
Bullseye	3.0	869	3.5	961
Red Dot	3.0	858	3.6	963
Unique	3.5	883	4.5	1039
Herco	3.6	840	4.6	1010

Accuracy Load:
Bullseye Powder, 3.0 Grains, 869 F.P.S.

NOTE: Loads shown in colored panels are maximum. These loads may be dangerous in your pistol, read preamble before using.

*Designates a compressed powder charge.

.38 SPECIAL

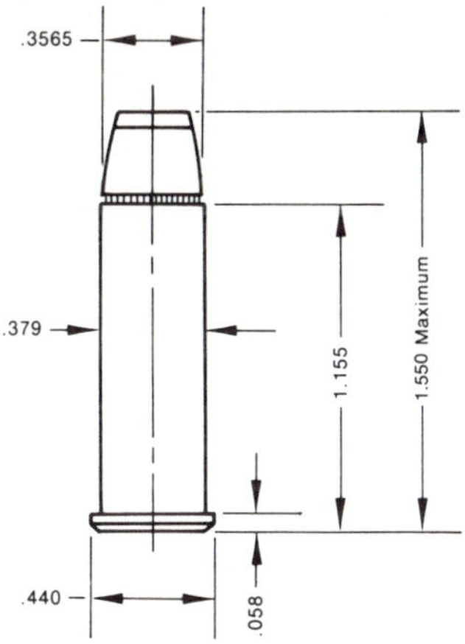

TEST COMPONENTS:

Cases .... Remington
Primers .... Remington 1½
Jacketed and Half-Jacketed Bullets Used ...
Speer 146 grain Half-Jacketed H.P.;
Remington 158 grain S.P. No. B22846;
Speer 160 grain Half-Jacketed

TEST (Velocity) SPECIFICATIONS:

Firearm Used . Smith and Wesson Model 14
Barrel Length .... 6''
Twist .... 1-18¾''
Groove Diameter .... .358''

NOTE: Handguns of various makes and models will occasionally vary in groove diameter. It is wise to slug your barrel before selecting a sizing die. Size cast bullets to as near actual groove diameter as possible.

To be seated correctly, cast bullet No. 358430 (in both 150 and 195 grain) must exceed the maximum overall length by .005''. All loads for this bullet take this extra length into consideration. We have used this bullet in its 195 grain weight for a 200 grain factory duplication load. This is due to the popularity of this particular bullet rather than an oversight.

Velocities for the half-jacketed bullets must be kept to at least 725 F.P.S. to prevent jackets from lodging in the barrel. If a crimp is desired on these half-jacketed bullets, it should be placed at the junction of the lead and the jacket.

For mid-range loads, we suggest bullet No. 358495 loaded at starting load velocities. Due to its light recoil, bullet No. 358101 is very popular for indoor (50 foot) target shooting.

Bullet No. 358311 duplicates the factory 158 grain service bullet very closely, while bullet No. 358429 is an excellent ''Keith type'' hunting bullet.

NOTE: Loads shown in colored panels are maximum. These loads may be dangerous in your pistol; read preamble before using.

*Designates a compressed powder charge.

.38 SPECIAL

146 Grain Half-Jacket

Powder	Sug. Starting Grains	Velocity F.P.S.	Max. Grains	Velocity F.P.S.
Unique	4.5	755	5.6	961
2400	9.5	790	11.0	987

158 Grain Jacketed

Powder	Sug. Starting Grains	Velocity F.P.S.	Max. Grains	Velocity F.P.S.
Unique	4.6	688	5.5	918
2400	9.5	785	11.0	960

160 Grain Half-Jacket

Powder	Sug. Starting Grains	Velocity F.P.S.	Max. Grains	Velocity F.P.S.
Unique	4.5	724	5.3	915
2400	9.5	845	10.8	985

112 Grain Cast
BULLET #358425 (#2 ALLOY)

Powder	Sug. Starting Grains	Velocity F.P.S.	Max. Grains	Velocity F.P.S.
Bullseye	2.0	572	3.8	954
Unique	3.5	682	6.9	1252
2400	8.0	656	12.5	1185
SR 7625	3.2	476	4.9	742
SR 4756	4.1	488	6.3	826
IMR 4227	7.0	416	9.9	757

Accuracy Load:
Bullseye Powder, 3.8 Grains, 954 F.P.S.

121 Grain Cast
BULLET #358242 (#2 ALLOY)
CAN ALSO USE BULLET #356402 (121 Grs.)

Powder	Sug. Starting Grains	Velocity F.P.S.	Max. Grains	Velocity F.P.S.
Bullseye	2.0	560	3.8	923
Unique	3.5	643	6.9	1132
2400	8.0	600	12.5	1120
SR 7625	3.2	465	4.9	786
SR 4756	4.1	450	6.3	821

Accuracy Load:
Unique Powder, 6.9 Grains, 1132 F.P.S.

141 Grain Cast
BULLET #358495 (#2 ALLOY)

Powder	Sug. Starting Grains	Velocity F.P.S.	Max. Grains	Velocity F.P.S.
Bullseye	2.0	568	3.7	937
Unique	3.5	739	6.8	1295
2400	8.0	794	12.0	1233
SR 7625	3.1	550	4.8	966
SR 4756	4.0	562	6.2	982
IMR 4227	7.0	550	9.7	833

Accuracy Load:
Unique Powder, 6.8 Grains, 1295 F.P.S.

150 Grain Cast
BULLET #358430 (#2 ALLOY)

Powder	Sug. Starting Grains	Velocity F.P.S.	Max. Grains	Velocity F.P.S.
Bullseye	2.0	485	3.6	811
Unique	3.5	608	5.5	940
2400	8.0	627	11.5	1015
SR 7625	3.5	456	4.7	730
SR 4756	4.5	471	6.1	732

Accuracy Load:
Bullseye Powder, 3.6 Grains, 811 F.P.S.

Factory Duplication Load:
2400 Powder, 11.5 Grains, 1015 F.P.S.

158 Grain Cast
BULLET #358311 (#2 ALLOY)
CAN ALSO USE BULLET #358156 (158 Grs. w/Gas Check)

Powder	Sug. Starting Grains	Velocity F.P.S.	Max. Grains	Velocity F.P.S.
Bullseye	2.0	519	3.5	826
Unique	3.5	660	5.4	1002
2400	8.0	690	11.0	1010
SR 7625	3.0	468	4.6	828
SR 4756	4.0	519	6.0	850
IMR 4227	7.0	504	9.5	755

Accuracy Load:
SR 4756 Powder, 6.0 Grains, 850 F.P.S.

Factory Duplication Load:
Bullseye Powder, 3.5 Grains, 826 F.P.S.

NOTE: Loads shown in colored panels are maximum. These loads may be dangerous in your pistol; read preamble before using. *Designates a compressed powder charge.

.357 MAGNUM

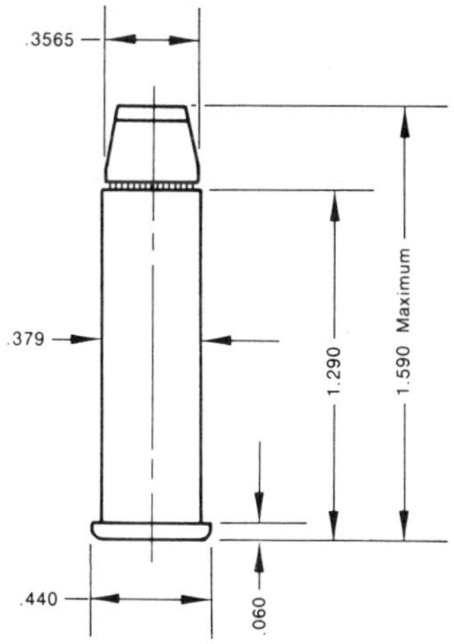

TEST COMPONENTS:

Cases .... Remington
Primers .... Remington 5½
Jacketed and Half-Jacketed Bullets Used ...
Speer 146 grain Half-Jacketed H.P.;
Remington 158 grain S.P. No. B22846;
Speer 160 grain Half-Jacketed

TEST (Velocity) SPECIFICATIONS:

Firearm Used . Smith and Wesson Model 27
Barrel Length .... 5''
Twist .... 1-18¾''
Groove Diameter .... .356''

NOTE: Do not use .38 Special cases for .357 Magnum loads. Use of the smaller case will increase pressure.

Depending upon their manufacture, some variations in groove diameter exist in these handguns. Slug your bore before reloading and size cast bullets to exact groove diameter.

In order to hold the maximum overall length to below 1.590'', it is sometimes necessary to crimp cast bullets on the forward edge of the first driving band.

Velocities for half-jacketed bullets must be kept above 750 F.P.S. to prevent jackets from lodging in the barrel.

For those desiring a mid-range load, we suggest bullet No. 358495 at starting load velocities. Bullet No. 358156 is extremely popular and an excellent choice for the heavier loads. Another great favorite is the ''Keith type'' bullet No. 358429.

NOTE: Loads shown in colored panels are maximum.

*Designates a compressed powder charge.

.357 MAGNUM Continued

146 Grain Half-Jacket

Powder	Sug. Starting Grains	Velocity F.P.S.	Max. Grains	Velocity F.P.S.
Unique	5.5	971	8.5	1300
2400	11.5	978	15.5	1323
SR 7625	6.1	845	6.5	971
SR 4756	7.0	858	9.0	1189
IMR 4227	11.5	830	15.5	1168

Accuracy Load:
Unique Powder, 8.5 Grains, 1300 F.P.S.

158 Grain Jacketed

Powder	Sug. Starting Grains	Velocity F.P.S.	Max. Grains	Velocity F.P.S.
Unique	5.5	801	8.0	1170
2400	11.5	918	15.0	1227
SR 7625	6.0	840	6.5	930
SR 4756	7.0	835	8.8	1108
IMR 4227	11.5	776	15.0	1128

Accuracy Load:
Unique Powder, 8.0 Grains, 1170 F.P.S.

160 Grain Half-Jacket

Powder	Sug. Starting Grains	Velocity F.P.S.	Max. Grains	Velocity F.P.S.
Unique	5.5	832	7.5	1152
2400	11.5	979	14.5	1221
SR 7625	5.9	890	6.4	960
SR 4756	7.0	835	8.7	1081
IMR 4227	11.5	868	14.8	1109

150 Grain Cast
BULLET #358430 (#2 ALLOY)

Powder	Sug. Starting Grains	Velocity F.P.S.	Max. Grains	Velocity F.P.S.
Bullseye	3.0	705	4.5	942
Unique	4.0	690	8.0	1280
Herco	7.0	1030	11.7*	1449
2400	11.0	934	15.0	1280
SR 7625	5.5	848	6.8	1082
SR 4756	6.0	754	9.6	1299
IMR 4227	9.5	657	15.5	1230

Accuracy Load:
SR 7625 Powder, 6.8 Grains, 1082 F.P.S.

158 Grain Cast (w/Gas Check)
BULLET #358156 (#2 ALLOY)
CAN ALSO USE BULLET #358311 (158 Grs.)

Powder	Sug. Starting Grains	Velocity F.P.S.	Max. Grains	Velocity F.P.S.
Bullseye	3.0	682	4.5	928
Unique	4.0	680	8.0	1250
Herco	7.0	1025	11.6*	1388
2400	11.0	956	15.0	1270
SR 7625	5.5	846	6.7	1064
SR 4756	6.0	712	9.5	1258
IMR 4227	9.5	725	15.0	1194

Accuracy Load:
2400 Powder, 15.0 Grains, 1270 F.P.S.

Factory Duplication Load:
Herco Powder, 11.6* Grains, 1388 F.P.S.

168 Grain Cast
BULLET #358429 (#2 ALLOY)

Powder	Sug. Starting Grains	Velocity F.P.S.	Max. Grains	Velocity F.P.S.
Bullseye	3.0	739	4.0	893
Unique	4.0	739	7.0	1157
Herco	7.0	1073	10.0*	1324
2400	11.0	1010	13.0	1180
SR 7625	5.0	896	6.0	1050
SR 4756	6.0	863	8.8	1182
IMR 4227	9.5	772	14.5	1212

Accuracy Load:
Unique Powder, 7.0 Grains, 1157 F.P.S.

195 Grain Cast
BULLET #358430 (#2 ALLOY)

Powder	Sug. Starting Grains	Velocity F.P.S.	Max. Grains	Velocity F.P.S.
Bullseye	3.0	720	3.5	782
Unique	4.0	719	5.5	942
Herco	6.0	944	7.5*	1097
2400	9.5	877	11.0	1018
SR 7625	4.2	763	5.2	894
SR 4756	5.5	813	7.2	1024
IMR 4227	9.5	792	12.2	1020

Accuracy Load:
Unique Powder, 5.5 Grains, 942 F.P.S.

NOTE: Loads shown in colored panels are maximum. These loads may be dangerous in your pistol; read preamble before using. *Designates a compressed powder charge.

.44 REMINGTON MAGNUM

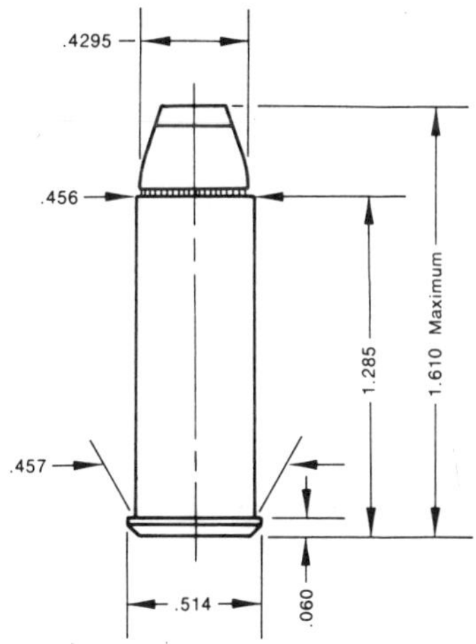

TEST COMPONENTS:

Cases .... Winchester
Primers .... Remington 2½
Jacketed and Half-Jacketed Bullets Used ...
Speer 225 grain, Half-Jacketed, H.P.;
Remington 240 grain S.P. No. B22906

TEST (Velocity) SPECIFICATIONS:

Firearm Used .. Smith & Wesson Model 29
Barrel Length .... 6½"
Twist .... 1-20"
Groove Diameter .... .4285"

NOTE: To prevent bullet movement when the gun is under recoil, the cartridge requires a hard crimp on all of the heavier loads.

Proper seating of some cast bullets requires that you exceed the maximum overall length listed. The correct overall lengths with these bullets are shown below. Data for these bullets takes this extra length into consideration.

Overall length with bullet No. 429303 - 1.692"
Overall length with bullet No. 429215 - 1.645"
Overall length with bullet No. 429360 - 1.660"
Overall length with bullet No. 429244 - 1.680"

Bullet No. 429348, loaded to starting load velocities, is recommended for mid-range use, while bullet No. 429215 is suggested as the best all round choice.

NOTE: Loads shown in colored panels are maximum.
*Designates a compressed powder charge.

.44 REMINGTON MAGNUM Continued

225 Grain Half Jacket

Powder	Sug. Starting Grains	Velocity F.P.S.	Max. Grains	Velocity F.P.S.
Unique	9.0	1035	13.0	1307
Herco	11.0	1087	14.5*	1296
2400	18.0	1090	23.0*	1434
SR 4756	12.0	1066	16.5*	1408
IMR 4227	21.0	1188	26.0*	1432
AL-8	15.0	1062	19.0*	1329

Accuracy Load:
IMR 4227 Powder, 26.0* Grains, 1432 F.P.S.

240 Grain Jacketed

Powder	Sug. Starting Grains	Velocity F.P.S.	Max. Grains	Velocity F.P.S.
Unique	9.0	974	12.0	1237
Herco	10.0	920	13.5*	1116
2400	18.0	1022	22.0*	1300
SR 4756	12.0	1042	15.5	1252
IMR 4227	20.0	991	25.0*	1331
AL-8	14.0	949	18.0*	1190

Accuracy Load:
IMR 4227 Powder, 25.0* Grains, 1331 F.P.S.

180 Grain Cast
BULLET #429348 (#2 ALLOY)

Powder	Sug. Starting Grains	Velocity F.P.S.	Max. Grains	Velocity F.P.S.
Bullseye	5.0	810	7.0	1003
Unique	8.0	918	13.0	1403
Herco	11.0	1094	15.0	1313
2400	18.0	990	24.0*	1495
SR 4756	12.0	1069	17.0	1517
IMR 4227	19.0	1012	27.0	1375
AL-8	14.0	1055	20.0	1390

Accuracy Load:
Unique Powder, 8.0 Grains, 918 F.P.S.

205 Grain Cast (w/Gas Check)
BULLET #429303 (#2 ALLOY)
CAN ALSO USE BULLET #42798

Powder	Sug. Starting Grains	Velocity F.P.S.	Max. Grains	Velocity F.P.S.
Bullseye	5.0	795	7.0	1031
Unique	9.0	1050	13.0	1420
Herco	11.0	1121	15.0	1357
2400	18.0	995	24.0*	1430
SR 4756	12.0	1152	17.0	1466
IMR 4227	19.0	988	27.0*	1472
AL-8	14.0	1088	20.0*	1388

Accuracy Load:
IMR 4227 Powder, 27.0* Grains, 1472 F.P.S.

215 Grain Cast (w/Gas Check)
BULLET #429215 (#2 ALLOY)

Powder	Sug. Starting Grains	Velocity F.P.S.	Max. Grains	Velocity F.P.S.
Bullseye	5.0	768	7.0	1004
Unique	9.0	1045	13.0	1364
Herco	11.0	1060	15.0	1316
2400	18.0	1065	24.0*	1420
SR 4756	12.0	1110	17.0	1418
IMR 4227	19.0	879	27.0*	1476
AL-8	14.0	1025	20.0*	1380

Accuracy Load:
Unique Powder, 9.0 Grains, 1045 F.P.S.

250 Grain Cast (w/Gas Check)
BULLET #429244 (#2 ALLOY)

Powder	Sug. Starting Grains	Velocity F.P.S.	Max. Grains	Velocity F.P.S.
Bullseye	4.0	647	6.0	858
Unique	7.0	819	11.0	1200
Herco	10.0	982	13.0	1160
2400	18.0	1088	22.0*	1295
SR 4756	11.0	1017	15.0	1284
IMR 4227	19.0	988	25.0*	1390
AL-8	13.0	882	18.0*	1245

Accuracy Load:
IMR 4227 Powder, 25.0* Grains, 1390 F.P.S.

NOTE: Loads shown in colored panels are maximum. These loads may be dangerous in your pistol; read preamble before using. *Designates a compressed powder charge.

.44 REMINGTON MAGNUM

232 Grain Cast
BULLET #429360 (#2 ALLOY)

Powder Grains	Sug. Starting F.P.S.	Velocity Grains	Max. F.P.S.	Velocity
Bullseye	4.0	687	6.5	962
Unique	9.0	1040	12.0	1330
Herco	10.0	1054	14.0	1284
2400	18.0	1095	23.0*	1379
SR 4756	12.0	1100	16.0	1393
IMR 4227	19.0	1022	26.0*	1459
AL-8	14.0	1055	19.0*	1345

Accuracy Load:
IMR 4227 Powder, 19.0 Grains, 1022 F.P.S.

NOTE: Loads shown in colored panels are maximum. These loads may be dangerous in your pistol; read preamble before using.

*Designates a compressed powder charge.

.45 A.C.P.

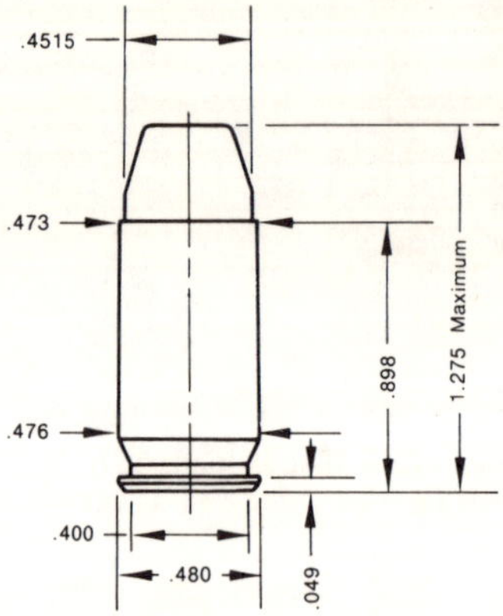

TEST COMPONENTS:

Cases .... Remington
Primers .... Remington 2½
Jacketed Bullets Used ....
Remington 185 grain, M.C. W.C. No. B22890;
Remington 230 grain M.C. No. B22892

TEST (Velocity) SPECIFICATIONS:

Firearm Used .... Colt, Government Model
Barrel Length .... 5''
Twist .... 1-16''
Groove Diameter .... .451''

NOTE: As this cartridge headspaces on the case mouth, the trim-to length must be adhered to closely and the case should not be crimped. A short case, or a crimp, can cause headspace problems.

While groove dimensions for these handguns will range from .450'' to .453'', the reloader may have difficulty if he uses a bullet over .451'' diameter. Case and chamber dimensions usually provide for a bullet of .451'', but larger bullets may bulge cases to the point where they fail to chamber.

Most pistols (unless they are altered) will not feed wad cutter type bullets such as No. 452389. Bullet No. 452460 feeds well in most pistols and, when loaded to starting load velocities, it is recommended for mid-range shooting.

NOTE: Loads shown in colored panels are maximum.
*Designates a compressed powder charge.

.45 A.C.P. Continued

185 Grain Jacketed

Powder	Sug. Starting Grains	Velocity F.P.S.	Max. Grains	Velocity F.P.S.
Bullseye	3.5	669	4.9	873
Red Dot	4.0	688	5.5	928
Unique	5.0	672	7.2	994
PB	4.8	762	5.8	900
SR 4756	6.2	606	6.6	664
SR 7625	5.2	639	6.2	840

Accuracy Load:
Unique Powder, 5.0 Grains, 672 F.P.S.

Factory Duplication Load:
Unique Powder, 5.9 Grains, 822 F.P.S.

230 Grain Jacketed

Powder	Sug. Starting Grains	Velocity F.P.S.	Max. Grains	Velocity F.P.S.
Bullseye	3.5	622	4.6	798
Red Dot	4.0	652	5.3	818
Unique	5.0	664	7.2	960
PB	4.5	650	5.6	835
SR 4756	6.1	636	6.5	690
SR 7625	5.1	673	6.1	878

Accuracy Load:
Bullseye Powder, 3.5 Grains, 622 F.P.S.

Factory Duplication Load:
Unique Powder, 6.5 Grains, 877 F.P.S.

185 Grain Cast
BULLET #452389 (#2 ALLOY)

Powder	Sug. Starting Grains	Velocity F.P.S.	Max. Grains	Velocity F.P.S.
Bullseye	3.5	778	4.9	966
Red Dot	4.0	805	5.5	1011
Unique	5.0	795	7.2	1081
PB	4.8	845	5.8	1000
SR 4756	6.3	755	6.6	793
SR 7625	5.2	778	6.2	1005

Accuracy Load:
Bullseye Powder, 4.9 Grains, 966 F.P.S.

200 Grain Cast
BULLET #452460 (#2 ALLOY)

Powder	Sug. Starting Grains	Velocity F.P.S.	Max. Grains	Velocity F.P.S.
Bullseye	3.5	731	4.8	913
Red Dot	4.0	769	5.4	959
Unique	5.0	752	7.2	1041
PB	4.5	738	5.8	964
SR 4756	6.2	715	6.5	737
SR 7625	5.2	763	6.2	864

Accuracy Load:
Bullseye Powder, 3.5 Grains, 731 F.P.S.

225 Grain Cast
BULLET #452374 (#2 ALLOY)

Powder	Sug. Starting Grains	Velocity F.P.S.	Max. Grains	Velocity F.P.S.
Bullseye	3.5	702	4.7	854
Red Dot	4.0	700	5.3	880
Unique	5.0	711	7.2	967
PB	4.5	655	5.7	840
SR 4756	6.2	638	6.5	653
SR 7625	5.1	655	6.1	827

Accuracy Load:
Bullseye Powder, 3.5 Grains, 702 F.P.S.

NOTE: Loads shown in colored panels are maximum.
*Designates a compressed powder charge.

Lyman Reloading Equipment

Presses

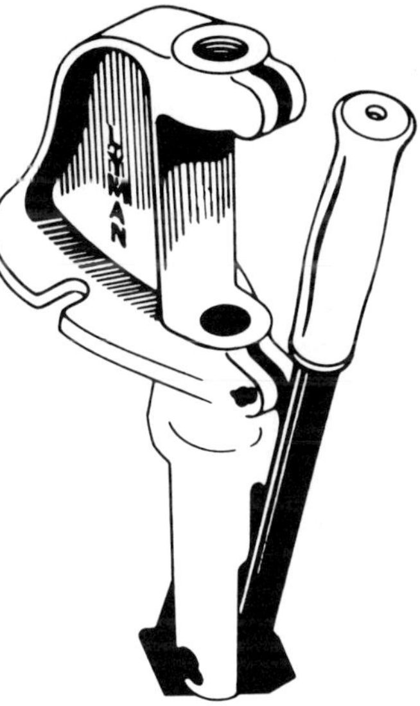

The Spartan

The Spartan Press reloads all rifle and pistol cases quickly, accurately and easily. Its powerful toggle-link mechanism multiplies the force applied to the handle 25 times and takes the hard work out of full-length resizing and case forming, and is rugged enough for bullet swaging. The Spartan 11 lb. iron frame reflects strength and quality throughout.

The Spar-T

The Lyman Spar-T Press combines the high speed of turret loading with the operating ease and strength of our ever-popular Spartan Press. The massive cast frame, ruggedly constructed of high-silicon iron-steel, firmly holds the 6 station Verti-Lock Turret in place with a heavy duty ¾'' steel stud. A positive stop, audible click action insures foolproof cartridge-to-die alignment and rapid operation.

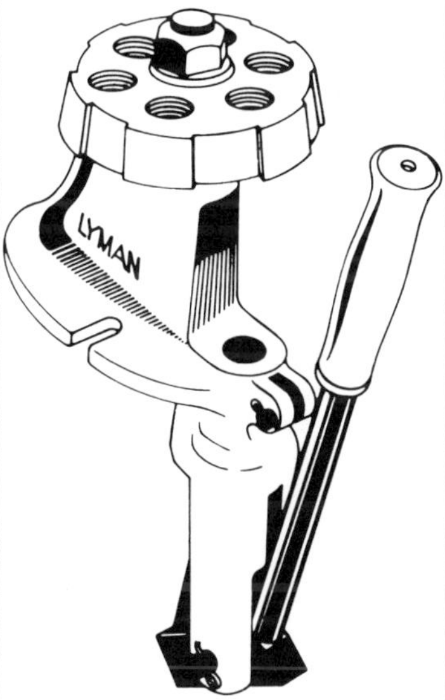

Presses

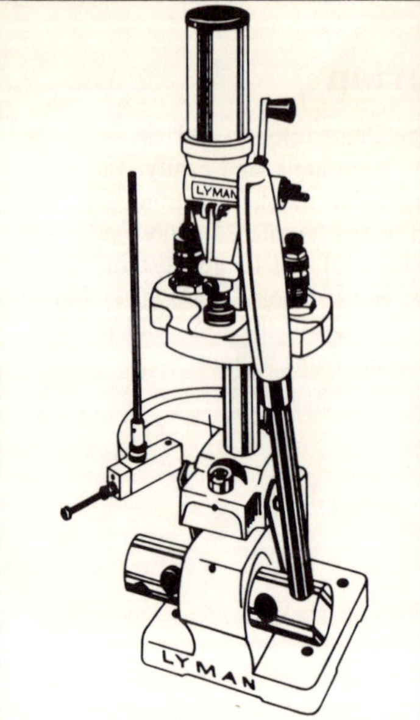

The All-American

The Lyman All-American Turret Press combines maximum speed with maximum precision. The four-station turret mounts three dies and a powder measure in reloading operation sequence. As the turret is revolved, dies remain in place, and are positioned directly over the shell holder with a positive-locking, audible-click action at each of the four turret stations. No single-station press can compare with the turret speed. Used with the Lyman No. 55 Powder Measure and the push-button primer feed, the All-American Turret Press handles extra high volume reloading.

No. 55 Powder Measure

The Lyman 55 Powder Measure dispenses charge after charge of smokeless powder with consistent accuracy to a fraction of a grain. The key to this unfailing precision is the 55's unique three-slide micrometer, adjustable cavity. Extra-fine adjustments of width and depth provide consistent charges with almost any powder. A 2400 grain reservoir of clear PVC plastic resists chemical action and protects the powder from harmful light rays.

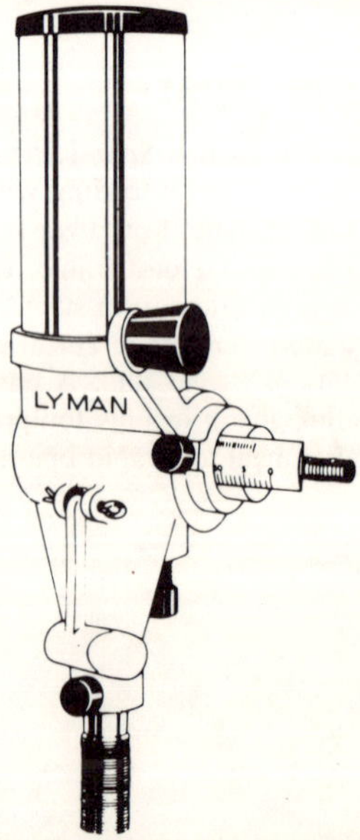

Reloading Dies

Reloading Dies

The Lyman All-American reloading die sets are designed for rapid, accurate reloading of metallic cartidges on our Spartan, Spar-T and All-American Turret presses, as well as all others with ⅞''x14 thread die stations. Carefully constructed of ''first quality'' steel, AA dies are first hardened and then polished to a mirror-smooth interior finish to produce a perfectly finished and sized case.

To assure the reloader years of service, the outer surfaces of the dies are chrome-plated for maximum protection. Precision rolled threads and hexagonal steel lock-rings securely hold die in precise adjustment during reloading. A handy die adjustment wrench comes with each set.

Deluxe Reloading Dies

The Lyman All-American Deluxe reloading die sets offer the ultimate in precise rifle and pistol cartridge reloading. The rifle sets feature the Lyman P-A seating and crimping die while the pistol sets include the special Lyman T-C (Tungsten Carbide) die for extra long life.

Accessories

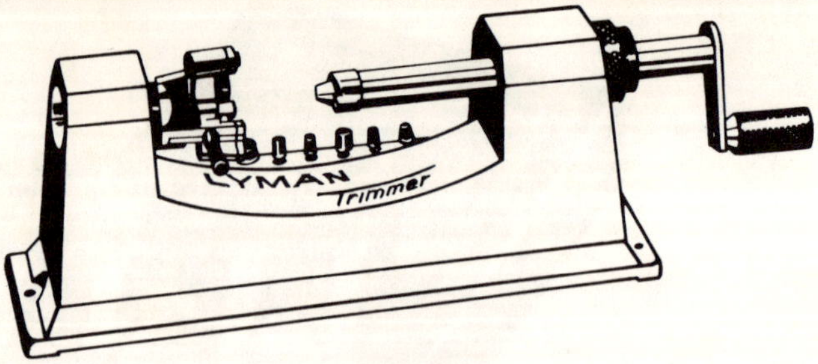

Lyman's Universal Trimmer

The Lyman Universal Trimmer features simplicity and precision. Its patented chuck head accepts all metallic rifle or pistol cases, regardless of rim thickness. To change calibers, simply change the inexpensive case head pilot. Other features include coarse and fine cutter adjustments and an oil impregnated bronze bearing. A rugged cast base assures precision alignment and years of service.

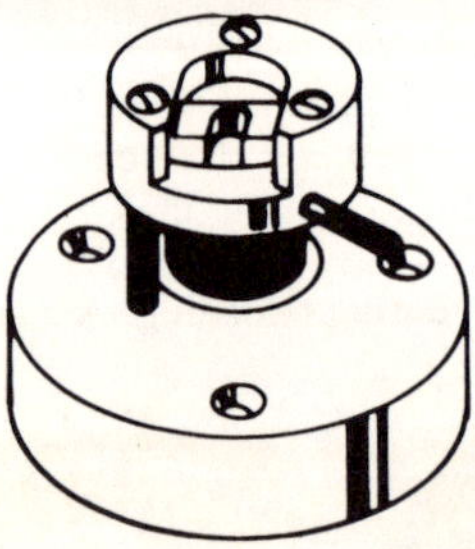

Universal Drill Press Trimmer

This Lyman universal chuck head bolts to your drill press. By mounting the cutter head and case head pilot to your drill chuck, you have a fast, accurate method of processing large quantities of cases.

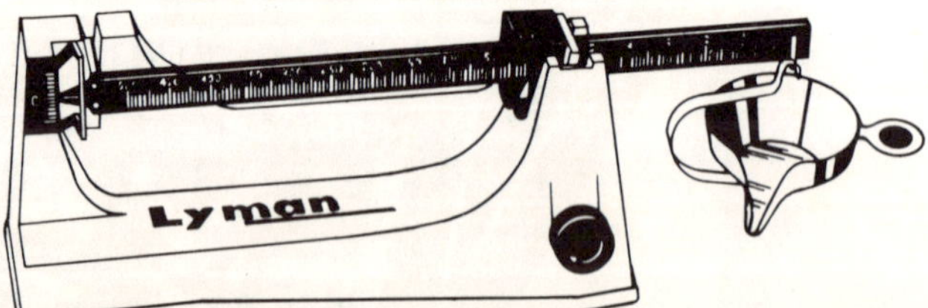

Lyman's D-7 Scale

The Lyman D-7 Scale represents the ultimate in a reloading scale. Its features add to up consistent, long-lasting accuracy.

A big 505 grain capacity will handle the heaviest charges and every scale is calibrated to a + 1/10 grain accuracy before shipment. A new, unique beam lifter guarantees less wear and tear when not in use or moving. Its magnetic damping and carbide knife edge on agate bearings assures the reloader of unsurpassed ease, speed and accuracy.

Other features include black on white beam markings, positive pan positioning and a heavy base casting for reduced vibration. The Lyman D-7 scale is the finest avialable.

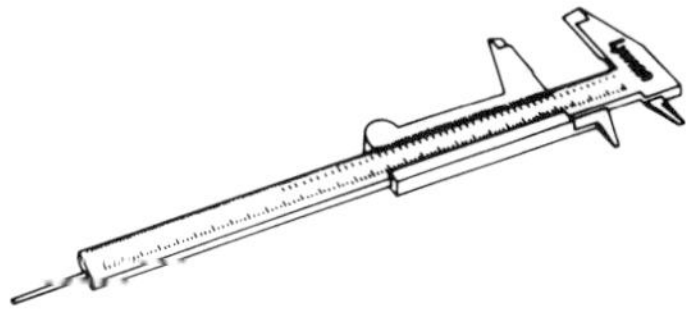

Lyman's Caliper

Lyman's caliper is accurate to + .015''. Direct inside, outside depth and stop measuring.

Lyman's Powder Dribbler

Lyman's powder dribbler permits use of the full measure of accuracy built into the powder scale. Features like a large powder reservoir, ideal height and tip-free, oversized base make this dribbler the ideal companion for any powder scale.

Lyman Burring Tool

Lyman's inside/outside burring tool is great for beveling and removing burrs from case mouths of new factory cases, formed cases and trimmed cases.

Case Lube Kit

A complete kit for efficient case lubrication. Includes Lyman's improved resizing lubricant, a ''fabric'' case lubricating pad and a handle with three interchangeable brushes covering every caliber.

Primer Pocket Reamer

Cleans and removes rough metal edges and carbon deposits from a primer pocket. This tool is a must for all reloaders.

Shell Chamfering Reamer

Tapers the case mouth for easier bullet seating. One size adapts to all cartridges, used with jacketed or cast bullets.

Powder Funnel

This Lyman plastic powder funnel is designed to fill cases from .22 Hornet through .45-70 without inserts or adjustments.

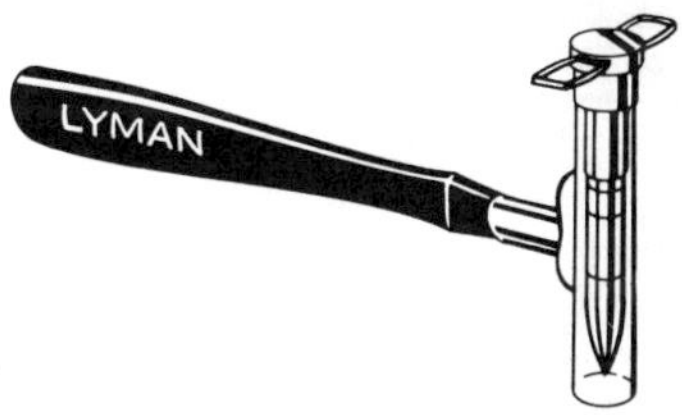

Inertia Bullet Puller

Quickly and easily removes bullets from cartridges.

Lyman's 45th Reloading Handbook

Here's the industry's biggest, most authoritative handbook, backed by our many years' experience filling the shooter's needs. This handbook is comprehensive enough for the advanced reloader, yet is clear enough to initiate the beginner. This complete handbook includes:

- 8,050 documented loads
- Covers 76 rifle calibers
- Covers 21 pistol calibers
- A complete muzzle loading section
- The "How-to" of reloading metallic cartridges covered in detail
- Complete specifics of all test components listed with data
- Plus a big reference section

Other great Lyman Publications:

Lyman Second Edition Shotshell Handbook

- Over 2,000 tested loads

Lyman Cast Bullet Handbook

- A must for every bullet caster

Lyman Black Powder Handbook

- The most comprehensive load information ever published

Lyman Wild Gourmet

- A Game Cookbook for the sportsman and gourmet alike

Lyman Fish Cookery

- The perfect companion to the Game Cookbook